# *Fundamentals of*
# Environmental Studies

*Fundamentals of*

# Environmental Studies

**Aditya Bhatnagar**

Oxford Book Company

Jaipur ▪ New Delhi

*Published By*

Oxford Book Company

**Regd. Off.** "Bony Residency", Gate No. 2
Opp. Tilak Public School, Vishweswariya Nagar,
Gopalpura Road, Jaipur - 302018, India
Ph: 0141 - 2761280, Telefax: 2761381
email - oxfordbookjpr@gmail.com
website - oxfordbookcompany.com

**Delhi Office:** 102, 1st Floor, Satyam House,
4327/3, Ansari Road, Darya Ganj,
New Delhi - 110002, India
Ph: 011-45652440

*Distributed By*
**KEYA INTERNATIONAL PRESS LLP**
E-mail: keyainternationalpressllp@gmail.com

Published in India, 2024



ISBN : 9789380179674



Typeset by Shivangi Computers, Jaipur

Printed and bound in India by Thomson Press Ltd.,
New Delhi

# Contents

# Contents

# Preface

Education about the environment and education through the environment is the only alternative to bring the compatibility between quality of life and quality of environment. For making environmental education, a successful movement, curriculum designers should think differently on the multidimensional tasks of developing curriculum with sustainable development as the intellectual centre.

The teachers should be equipped with pedagogical skills to transact the multifaceted curriculum encompassing cognitive, social, emotional and ethical dimension. Considering the above view point, an effort has been made in this book to address the emerging trends in environmental education. The teachers and students of all discipline in general and teacher education in particular will find this book useful.

*Aditya Bhatnagar*

# Chapter 1

# Facts on Biodiversity

Biodiversity is the measure of the number, variety and variability of living organisms. It includes diversity within species, between species, and among ecosystems. The concept also covers how this diversity changes from one location to another and over time. Indicators such as the number of species in a given area can help in monitoring certain aspects of biodiversity. Biodiversity is everywhere, both on land and in water. It includes all organisms, from microscopic bacteria to more complex plants and animals.

Current inventories of species, though useful, remain incomplete and insufficient for providing an accurate picture of the extent and distribution of all components of biodiversity. Based on present knowledge of how biodiversity changes over time, rough estimates can be made of the rates at which species become extinct. Ecosystem services are the benefits people obtain from ecosystems.

Biodiversity plays an important role in the way ecosystems function and in the many services they provide. Services include nutrient and water cycling, soil formation and retention, resistance against invasive species, pollination of plants, regulation of climate, as well as pest and pollution control by ecosystems. For ecosystem services it matters which species are abundant as well as how many species are present.

## KEY BENEFITS TO BIODIVERSITY

Biodiversity provides many key benefits to humans that go beyond the mere provision of raw materials. Biodiversity loss has negative effects on several aspects of human wellbeing, such as food security, vulnerability to natural disasters, energy security, and access to clean water and raw materials. It also affects human health, social relations, and freedom of choice. Society tends to have various competing goals, many of which depend on biodiversity.

When humans modify an ecosystem to improve a service it provides, this generally also results in changes to other ecosystem services. For example, actions to increase food production can lead to reduced water availability for other uses. As a result of such trade-offs, many services have been degraded, for instance fisheries, water supply, and protection against natural hazards. In the long term, the value of services lost may greatly exceed the short-term economic benefits that are gained from transforming ecosystems. Unlike goods bought and sold in markets, many ecosystem services are not traded in markets for readily observable prices.

This means that the importance of biodiversity and natural processes in providing benefits to humans is ignored by financial markets. New methods are being used to assign monetary values to benefits such as recreation or clean drinking water. Degradation of ecosystem services could be significantly slowed down or reversed if the full economic value of these services were taken into account in decision-making.

Over the last century, some people have benefited from the conversion of natural ecosystems and an increase in international trade, but other people have suffered from the consequences of biodiversity losses and from restricted access to resources they depend upon. Changes in ecosystems are harming many of the world's poorest people, who are the least able to adjust to these changes.

## CURRENT TRENDS IN BIODIVERSITY

Virtually all of Earth's ecosystems have been dramatically transformed through human actions and ecosystems continue to be converted for agricultural and other uses. The current loss of biodiversity and the related changes in the environment are now faster than ever before in human history and there is no sign of this process slowing down.

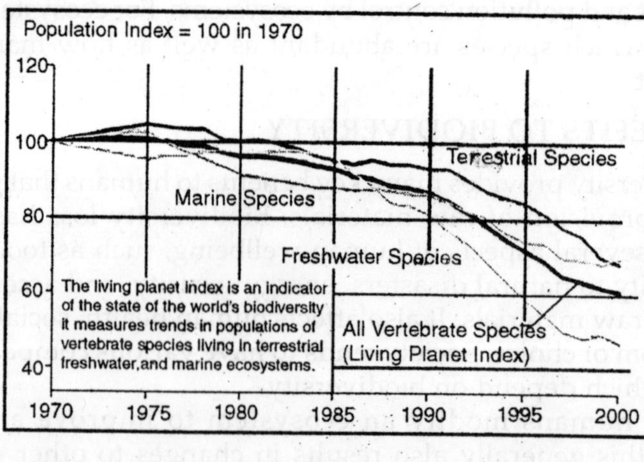

Many animal and plant populations have declined in numbers, geographical spread, or both. Species extinction is a natural part of Earth's history. Human activity has increased the extinction rate by at least 100 times compared to the natural rate.

Comparing different types of measurements of biodiversity loss is not simple. The rate of change in one aspect of biodiversity, such as loss of species richness, does not necessarily reflect the change in another, such as habitat loss.

Moreover, some aspects of biodiversity loss are not easily measured, for instance the fact that the same species are increasingly found at different locations on the planet and that overall biodiversity is decreasing.

## Factors Lead to Biodiversity Loss

Biodiversity is declining rapidly due to factors such as land use change, climate change, invasive species, over exploitation, and pollution. Such natural or human-induced factors – referred to as drivers – tend to interact and amplify each other. While changes in biodiversity are more clearly linked to direct drivers such as habitat loss, they are also linked to indirect drivers that are at the root of many changes in ecosystems.

The main indirect drivers are changes in human population, economic activity, and technology, as well as sociopolitical and cultural factors. Different direct drivers have been critically important in different ecosystems over the past 50 years. For example, in terrestrial ecosystems, the main driver has been land cover change such as the conversion of forest to agriculture. In marine systems, however, fishing, and particularly over fishing, have been the main drivers of biodiversity loss.

Overall, the main factors directly driving biodiversity loss are: habitat change, such as fragmentation of forests; invasive alien species that establish and spread outside their normal distribution; over exploitation of natural resources; and pollution, particularly by excessive fertilizer use leading to excessive levels of nutrients in soil and water. Recent changes in climate have already had significant impacts on biodiversity and ecosystems in certain regions. As climate change will become more severe, the harmful impacts on ecosystem services are expected to outweigh possible benefits, such as a longer growing season, in most regions of the world. Climate change is expected to exacerbate risks of extinctions, floods, droughts, population declines, and disease outbreaks.

Many drivers affecting biodiversity are stronger today than they were in the past and are also occurring together. Because exposure to one threat often makes a species more susceptible to another, multiple threats may have unexpectedly dramatic impacts on biodiversity. Drivers of extinction range from local to global in scope and from immediate to long-term in their effects. For example, the extinction of species due to habitat loss can be rapid for some species, while it may take hundreds of years for others.

## How Might Biodiversity Change in the Future under Various Plausible Scenarios?

The Millennium Ecosystem Assessment developed four plausible scenarios to explore the future of biodiversity and human well-being until 2050 and beyond. The different scenarios are based on either increased globalization or increased regionalization, and an either reactive or proactive way of addressing environmental issues.

Overall, in all four scenarios, agricultural land will expand and forest cover will shrink, particularly in developing countries. This will lead to a continuing decline in local and global biodiversity, mainly as a result of habitat loss. More proactive approaches to the environment will be more successful in slowing these trends. Aquatic biodiversity and specific fish populations are expected to decline due to factors such as excessive levels of nutrients, over harvesting, invasion by alien species, and pollution.

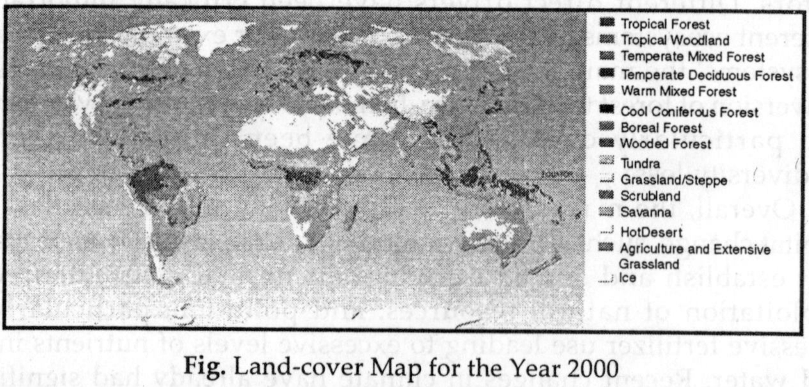


**Fig.** Land-cover Map for the Year 2000

Human well-being will be affected by biodiversity loss both directly and indirectly. Direct effects include an increased risk of sudden environmental changes such as fisheries collapses, floods, droughts, wildfires, and disease. Changes will also affect human well-being indirectly, for instance in the form of conflicts due to scarcer food and water resources.

Though the average income per person is projected to rise in all scenarios, this can mask increased inequity for instance in terms of food security. Major decisions will have to address trade-offs between competing goals, for instance between agricultural production and water quality, or between water use and aquatic biodiversity. Policies that conserve more biodiversity are also promoting higher overall human well-being by preserving multiple benefits obtained from ecosystems.

## ACTIONS TO CONSERVE BIODIVERSITY

Protected areas are an essential part of conservation programs, but they are not sufficient by themselves to protect the full range of biodiversity and can be difficult to enforce. To be successful, sites for protected areas need to be carefully chosen, ensuring that all regional ecosystems are well represented, and the areas need to be well designed and effectively managed.

Market tools, such as direct payments for ecosystem services or transfers of ownership rights to private individuals, can provide economic incentives to conserve biodiversity and to use ecosystem services sustainable. Prevention and early intervention have proven to be the most successful and cost-effective way of tackling invasive species. Once an invasive species has become established, its control and particularly its eradication through the use of chemicals or through the introduction of other species is not necessarily effective and is extremely difficult and costly.

To be conserved, biodiversity must be integrated into the agriculture, fishery, and forestry sectors. These sectors are directly dependent on biodiversity and affect it directly. The private sector can make significant contributions, for example by adopting certain agricultural practices. Many companies now show greater corporate responsibility and are preparing their own biodiversity action plans.

Strong institutions at all levels are essential to support biodiversity conservation and the sustainable use of ecosystems. International agreements need to include enforcement measures and take into account impacts on biodiversity and possible synergies with other agreements. Most direct actions to halt or reduce biodiversity loss need to be taken at local or national level.

Suitable laws and policies developed by central governments can enable local levels of government to provide incentives for sustainable resource management. Informing all of society about the benefits of conserving biodiversity, and explicitly considering trade-offs between different options in an integrated way, helps maximize the benefits to society.

Ecosystem restoration is generally far more expensive than protecting the original ecosystem, but is becoming increasingly important as more areas become degraded. Direct and indirect drivers of biodiversity loss must be addressed to better protect biodiversity and ecosystem services.

Possible actions include eliminating harmful subsidies, promoting sustainable intensification of agriculture, adapting to climate change, limiting the increase in nutrient levels in soil and water, assessing the full economic value of ecosystem services, and increasing the transparency of decision making processes.

Fig. Biodiversity Includes all Organisms, from Bacteria to Complex Animals. Biodiversity Provides

In 2002, the Parties to the Convention on Biological Diversity agreed on a target to achieve a "significant reduction of the current rate of biodiversity loss at the global, regional, and national level as a contribution to poverty alleviation and to the benefit of all life on earth" by 2010. Given appropriate actions, it is possible to achieve a reduction in the rate of biodiversity loss for certain components of biodiversity and in certain regions within that time frame.

However, a reduction in the overall rate of biodiversity loss is unlikely to be achieved by 2010. Indeed, current trends show no sign of a slowdown of biodiversity loss, and direct drivers of loss such as land use change and climate change are expected to increase further. Moreover, it can take many years for institutions to take actions and for the positive and negative impacts of human actions on biodiversity and ecosystems to become apparent.

Since changes take place over different time frames, longer-term goals and targets—say, for 2050—are needed to guide policy and actions, in addition to short-term targets. Even on economic grounds alone, there is substantial scope for greater protection of biodiversity. Ultimately, however, the level of biodiversity that survives on Earth will be determined not just by considerations of usefulness but also by ethical concerns. Tradeoffs between promoting human well-being and limiting biodiversity loss are likely, but synergies are also possible.

## VARIATION OF LIFE

Biodiversity is the variation of life forms within a given ecosystem, biome, or for the entire Earth. Biodiversity is often used as a measure of the health of biological systems. The biodiversity found on Earth today consists of many millions of distinct biological species, which is the product of nearly 3.5 billion years of evolution.

## ETYMOLOGY

Biodiversity is a portmanteau word, from biology and diversity, originating from and used interchangeably with "biological diversity." This term was used first by wildlife scientist and conservationist Raymond F. Dasmann in a lay book advocating nature conservation. The term was not widely adopted for more than a decade, when in the 1980s it and "biodiversity" came into common usage in science and environmental policy. Use of the term by Thomas Lovejoy in the Foreword to the book credited with launching the field of conservation biology introduced the term along with "conservation biology" to the scientific community.

Until then the term "natural diversity" was used in conservation science circles, including by The Science Division of The Nature Conservancy in an important 1975 study, "The Preservation of Natural Diversity." By the early 1980s TNC's Science programme and its head Robert E. Jenkins, Lovejoy, and other leading conservation scientists at the time in America advocated the use of "biological diversity" to embrace the object of biological conservation.

The term's contracted form *biodiversity* may have been coined by W.G. Rosen in 1985 while planning the *National Forum on Biological Diversity* organized by the National Research Council (NRC) which was to be held in 1986, and first appeared in a publication in 1988 when entomologist E. O. Wilson used it as the title of the proceedings of that forum.

Since this period both terms and the concept have achieved widespread use among biologists, environmentalists, political leaders, and concerned citizens worldwide. The term is sometimes used to equate to a concern for the natural environment and nature conservation. This use has coincided with the expansion of concern over extinction observed in the last decades of the 20th century.

A similar concept in use in the United States, besides natural diversity, is the term "natural heritage." It predates both terms though it is a less scientific term and more easily comprehended in some ways by the wider audience interested in conservation. "Natural Heritage" was used when Jimmy Carter set up the Georgia Heritage Trust while he was governor of Georgia; Carter's trust dealt with both natural and cultural heritage.

It would appear that Carter picked the term up from Lyndon Johnson, who used it in a 1966 Message to Congress. "Natural Heritage" was picked up by the Science Division of the US Nature Conservancy when, under Jenkins, it launched in 1974 the network of State Natural Heritage Programs. When this network was extended outside the USA, the term "Conservation Data Centre" was suggested by Guillermo Mann and came to be preferred.

Biologists most often define "biological diversity" or "biodiversity" as the "totality of genes, species, and ecosystems of a region". An advantage of this definition is that it seems to describe most circumstances and present a unified view of the traditional three levels at which biological variety has been identified:

- Genetic diversity
- Species diversity
- Ecosystem diversity

This multilevel conception is consistent with the early use of "biological diversity" in Washington. D.C. and international conservation organizations in the late 1960s through 1970's, by Raymond F. Dasmann who apparently coined the term and Thomas E. Lovejoy who later introduced it to the wider conservation and science communities.

An explicit definition consistent with this interpretation was first given in a paper by Bruce A. Wilcox commissioned by the International Union for the Conservation of Nature and Natural Resources (IUCN) for the 1982 World National Parks Conference in Bali.

The definition Wilcox gave is "Biological diversity is the variety of life forms...at all levels of biological systems (i.e., molecular, organismic, population, species and ecosystem)..." Subsequently, the 1992 United Nations Earth Summit in Rio de Janeiro defined "biological diversity" as "the variability among living organisms from all sources, including, 'inter alia', terrestrial, marine, and other aquatic ecosystems, and the ecological complexes of which they are part: this includes diversity within species, between species and of ecosystems". This is, in fact, the closest thing to a single legally accepted definition of biodiversity, since it is the definition adopted by the United Nations Convention on Biological Diversity.

The current textbook definition of "biodiversity" is "variation of life at all levels of biological organization". For geneticists, *biodiversity* is the diversity of genes and organisms. They study processes such as mutations, gene exchanges, and genome dynamics that occur at the DNA level and generate evolution. Consistent with this, along with the above definition the Wilcox paper stated "genes are the ultimate source of biological organization at all levels of biological systems..."

## MEASUREMENT

A variety of objective measures have been created in order to empirically measure biodiversity. Each measure of biodiversity relates to a particular use of the data. For practical conservationists, measurements should include a quantification of values that are commonly-shared among locally affected organisms, including humans. For others, a more economically defensible definition should allow the ensuring of continued possibilities for both adaptation and future use by humans, assuring environmental sustainability.

As a consequence, biologists argue that this measure is likely to be associated with the variety of genes. Since it cannot always be said which genes are more likely to prove beneficial, the best choice for conservation is to assure the persistence of as many genes as possible. For ecologists, this latter approach is sometimes considered too restrictive, as it prohibits ecological succession.

Biodiversity is usually plotted as taxonomic richness of a geographic area, with some reference to a temporal scale. Whittaker described three common metrics used to measure species-level biodiversity, encompassing attention to species richness or species evenness:

- Species richness - the least sophisticated of the indices available.
- Simpson index.
- Shannon-Wiener index.

Recently, another new index has been invented called the Mean Species Abundance Index (MSA); this index calculates the trend in population size of a cross section of the species. It does this in line with the CBD 2010 indicator for species abundance.

There are three other indices which are used by ecologists:

- Alpha diversity refers to diversity within a particular area, community or ecosystem, and is measured by counting the number of taxa within the ecosystem (usually species)
- Beta diversity is species diversity between ecosystems; this involves comparing the number of taxa that are unique to each of the ecosystems.

- Gamma diversity is a measurement of the overall diversity for different ecosystems within a region.

## DISTRIBUTION

Selection bias continues to bedevil modern estimates of biodiversity. In 1768 Rev. Gilbert White succinctly observed of his Selborne, Hampshire "all nature is so full, that that district produces the most variety which is the most examined."

Nevertheless, biodiversity is not distributed evenly on Earth. It is consistently richer in the tropics and in other localized regions such as the Cape Floristic Province. As one approaches polar regions one generally finds fewer species. Flora and fauna diversity depends on climate, altitude, soils and the presence of other species. In the year 2006 large numbers of the Earth's species were formally classified as rare or endangered or threatened species; moreover, many scientists have estimated that there are millions more species actually endangered which have not yet been formally recognized. About 40 percent of the 40,177 species assessed using the IUCN Red List criteria, are now listed as threatened species with extinction - a total of 16,119 species.

Even though biodiversity declines from the equator to the poles in terrestrial ecoregions, whether this is so in aquatic ecosystems is still a hypothesis to be tested, especially in marine ecosystems where causes of this phenomenon are unclear. In addition, particularly in marine ecosystems, there are several well stated cases where diversity in higher latitudes actually increases. Therefore, the lack of information on biodiversity of Tropics and Polar Regions prevents scientific conclusions on the distribution of the world's aquatic biodiversity.

A biodiversity hotspot is a region with a high level of endemic species. These biodiversity hotspots were first identified by Dr. Norman Myers in two articles in the scientific journal *The Environmentalist*. Dense human habitation tends to occur near hotspots. Most hotspots are located in the tropics and most of them are forests.

Brazil's Atlantic Forest is considered a hotspot of biodiversity and contains roughly 20,000 plant species, 1350 vertebrates, and millions of insects, about half of which occur nowhere else in the world. The island of Madagascar including the unique Madagascar dry deciduous forests and lowland rainforests possess a very high ratio of species endemism and biodiversity, since the island separated from mainland Africa 65 million years ago, most of the species and ecosystems have evolved independently producing unique species different from those in other parts of Africa.

Many regions of high biodiversity (as well as high endemism) arise from very specialized habitats which require unusual adaptation mechanisms. For example the peat bogs of Northern Europe.

## EVOLUTION

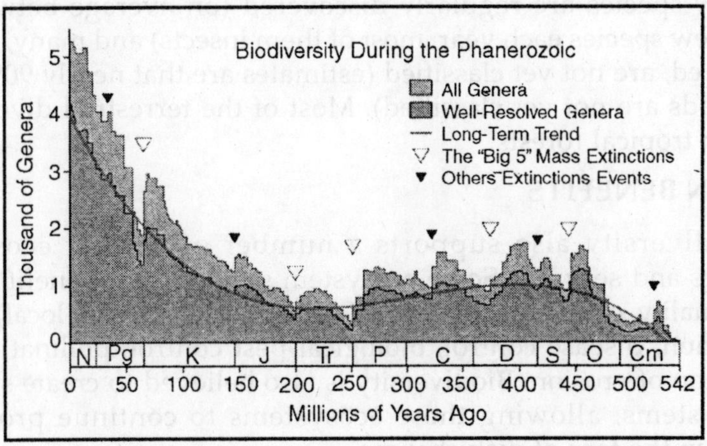


**Fig.** Apparent Marine Fossil Diversity During the Phanerozoic Eon

Biodiversity found on Earth today is the result of 4 billion years of evolution. The origin of life has not been definitely established by science, however some evidence suggests that life may already have been well-established a few hundred million years after the formation of the Earth. Until approximately 600 million years ago, all life consisted of archaea, bacteria, protozoans and similar single-celled organisms.

The history of biodiversity during the Phanerozoic (the last 540 million years), starts with rapid growth during the Cambrian explosion—a period during which nearly every phylum of multicellular organisms first appeared. Over the next 400 million years or so, global diversity showed little overall trend, but was marked by periodic, massive losses of diversity classified as mass extinction events.

The apparent biodiversity shown in the fossil record suggests that the last few million years include the period of greatest biodiversity in the Earth's history. However, not all scientists support this view, since there is considerable uncertainty as to how strongly the fossil record is biased by the greater availability and preservation of recent geologic sections. Some argue that, corrected for sampling artifacts, modern biodiversity is not much different from biodiversity 300 million years ago. Estimates of the present global macroscopic species diversity vary from 2 million to 100 million species, with a best estimate of somewhere near 13–14 million, the vast majority of them arthropods.

Most biologists agree however that the period since the emergence of humans is part of a new mass extinction, the Holocene extinction event, caused primarily by the impact humans are having on the environment. It has been argued that the present rate of extinction is sufficient to eliminate most species on the planet Earth within 100 years.

New species are regularly discovered (on average between 5–10,000 new species each year, most of them insects) and many, though discovered, are not yet classified (estimates are that nearly 90% of all arthropods are not yet classified). Most of the terrestrial diversity is found in tropical forests.

## HUMAN BENEFITS

Biodiversity also supports a number of natural ecosystem processes and services. Some ecosystem services that benefit society are air quality, climate (both global $CO_2$ sequestration and local), water purification, disease control, biological pest control, pollination and prevention of erosion. Biodiversity is also believed to create stability in ecosystems, allowing these ecosystems to continue providing services in the face of disturbances.

Non-material benefits that are obtained from ecosystems include spiritual and aesthetic values, knowledge systems and the value of education. Biodiversity is also central to an ecocentric philosophy.

### Agriculture

The economic value of the reservoir of genetic traits present in wild varieties and traditionally grown landraces is extremely important in improving crop performance.

Important crops, such as the potato and coffee, are often derived from only a few genetic strains. Improvements in crop plants over the last 250 years have been largely due to harnessing the genetic diversity present in wild and domestic crop plants.

Interbreeding crops strains with different beneficial traits has resulted in more than doubling crop production in the last 50 years as a result of the Green Revolution.

Crop diversity is also necessary to help the system recover when the dominant crop type is attacked by a disease:

- The Irish potato blight of 1846, which was a major factor in the deaths of a million people and migration of another million, was the result of planting only two potato varieties, both of which were vulnerable.
- When rice grassy stunt virus struck rice fields from Indonesia to India in the 1970s. 6273 varieties were tested for resistance.

One was found to be resistant, an Indian variety, known to science only since 1966. This variety formed a hybrid with other varieties and is now widely grown.

- Coffee rust attacked coffee plantations in Sri Lanka, Brazil, and Central America in 1970. A resistant variety was found in Ethiopia.

Although the diseases are themselves a form of biodiversity.

Monoculture, the lack of biodiversity, was a contributing factor to several agricultural disasters in history, including the Irish Potato Famine, the European wine industry collapse in the late 1800s, and the US Southern Corn Leaf Blight epidemic of 1970.

Higher biodiversity also controls the spread of certain diseases as pathogens will need to adapt to infect different species.

Biodiversity provides food for humans. Although about 80 percent of our food supply comes from just 20 kinds of plants, humans use at least 40,000 species of plants and animals a day. Many people around the world depend on these species for their food, shelter, and clothing. There is untapped potential for increasing the range of food products suitable for human consumption, provided that the high present extinction rate can be stopped.

**Human Health**

The relevance of biodiversity to human health is becoming a major international political issue, as scientific evidence builds on the global health implications of biodiversity loss.

This issue is closely linked with the issue of climate change, as many of the anticipated health risks of climate change are associated with changes in biodiversity (e.g. changes in populations and distribution of disease vectors, scarcity of fresh water, impacts on agricultural biodiversity and food resources etc).

Some of the health issues influenced by biodiversity include dietary health and nutrition security, infectious diseases, medical science and medicinal resources, social and psychological health, and spiritual well-being. Biodiversity is also known to have an important role in reducing disaster risk, and in post-disaster relief and recovery efforts.

One of the key health issues associated with biodiversity is that of drug discovery and the availability of medicinal resources. A significant proportion of drugs are derived, directly or indirectly, from biological sources; Chivian and Bernstein report that at least 50% of the pharmaceutical compounds on the market in the US are derived from natural compounds found in plants, animals, and microorganisms, while about 80% of the world population depends on medicines from nature (used in either modern or traditional medical practice) for primary healthcare.

Moreover, only a tiny proportion of the total diversity of wild species has been investigated for potential sources of new drugs. Through the field of bionics, considerable technological advancement has occurred which would not have without a rich biodiversity. It has been argued, based on evidence from market analysis and biodiversity science, that the decline in output from the pharmaceutical sector since the mid-1980s can be attributed to a move away from natural product exploration ("bioprospecting") in favour of R&D programmes based on genomics and synthetic chemistry, neither of which have yielded the expected product outputs; meanwhile, there is evidence that natural product chemistry can provide the basis for innovation which can yield significant economic and health benefits. Marine ecosystems are of particular interest in this regard, however unregulated and inappropriate bioprospecting can be considered a form of over-exploitation which has the potential to degrade ecosystems and increase biodiversity loss, as well as impacting on the rights of the communities and states from which the resources are taken.

## Business and Industry

A wide range of industrial materials are derived directly from biological resources. These include building materials, fibers, dyes, resirubber and oil. There is enormous potential for further research into sustainably utilizing materials from a wider diversity of organisms. In addition, biodivesity and the ecosystem goods and services it provides are considered to be fundamental to healthy economic systems.

The degree to which biodiversity supports business varies between regions and between economic sectors, however the importance of biodiversity to issues of resource security (water quantity and quality, timber, paper and fibre, food and medicinal resources etc) are increasingly recognized as universal. As a result, the loss of biodiversity is increasingly recognized as a significant risk factor in business development and a threat to long term economic sustainability. A number of case studies recently compiled by the World Resources Institute demonstrate some of these risks as identified by specific industries.

## Other Ecological Services

Biodiversity provides many ecosystem services that are often not readily visible. It plays a part in regulating the chemistry of our atmosphere and water supply.

Biodiversity is directly involved in water purification, recycling nutrients and providing fertile soils. Experiments with controlled environments have shown that humans cannot easily build ecosystems to support human needs; for example insect pollination cannot be mimicked by human-made construction, and that activity alone represents tens of billions of dollars in ecosystem services per annum to humankind.

The stability of ecosystems is also related to biodiversity, with higher biodiversity producing greater stability over time, reducing the chance that ecosystem services will be disrupted as a result of disturbances such as extreme weather events or human exploitation.

**Leisure, Cultural and Aesthetic Value**

Many people derive value from biodiversity through leisure activities such as hiking, birdwatching or natural history study. Biodiversity has inspired musicians, painters, sculptors, writers and other artists.

Many cultural groups view themselves as an integral part of the natural world and show respect for other living organisms.

Popular activities such as gardening, caring for aquariums and collecting butterflies are all strongly dependent on biodiversity.

The number of species involved in such pursuits is in the tens of thousands, though the great majority do not enter mainstream commercialism.

The relationships between the original natural areas of these often 'exotic' animals and plants and commercial collectors, suppliers, breeders, propagators and those who promote their understanding and enjoyment are complex and poorly understood. It seems clear, however, that the general public responds well to exposure to rare and unusual organisms—they recognize their inherent value at some level. A family outing to the botanical garden or zoo is as much an aesthetic or cultural experience as it is an educational one.

Philosophically it could be argued that biodiversity has intrinsic aesthetic and spiritual value to mankind *in and of itself*. This idea can be used as a counterweight to the notion that tropical forests and other ecological realms are only worthy of conservation because they may contain medicines or useful products. An interesting point is that evolved DNA embodies knowledge, and therefore destroying a species resembles burning a book, with the caveat that the book is of uncertain depth and importance and may in fact be best used as fuel.

## Number of Species

*As a general guide, the numbers of identified modern species as of 2004 can be broken down as follows:*

| | |
|---|---|
| • 287,655 plants, including:<br>– 15,000 mosses,<br>– 13,025 ferns,<br>– 980 gymnosperms,<br>– 199,350 dicotyledons,<br>– 59,300 monocotyledons;<br>• 74,000–120,000 fungi;<br>• 10,000 lichens; | • 1,250,000 animals, including:<br>– 1,190,200 invertebrates:<br>• 950,000 insects,<br>• 70,000 mollusks,<br>• 40,000 crustaceans,<br>• 130,200 others;<br>– 58,808 vertebrates:<br>• 29,300 fish,<br>• 5,743 amphibians,<br>• 8,240 reptiles,<br>• 10,234 birds, (9799 extant as of 2006)<br>• 5,416 mammals. |

*However the total number of species for some phyla may be much higher:*

- 10–30 million insects.
- 5–10 million bacteria.
- 1.5 million fungi.
- ~1 million mites.

## THREATS

During the last century, erosion of biodiversity has been increasingly observed. Some studies show that about one eighth of known plant species are threatened with extinction. Some estimates put the loss at up to 140,000 species per year (based on Species-area theory) and subject to discussion. This figure indicates unsustainable ecological practices, because only a small number of species come into being each year. Almost all scientists acknowledge that the rate of species loss is greater now than at any time in human history, with extinctions occurring at rates hundreds of times higher than background extinction rates.

The factors that threaten biodiversity have been variously categorized. Jared Diamond describes an "Evil Quartet" of habitat destruction, overkill, introduced species, and secondary extensions. Edward O. Wilson prefers the acronym HIPPO, standing for Habitat destruction, Invasive species, Pollution, Human Over Population, and Overharvesting.

The most authoritative classification in use today is that of IUCN's

Classification of Direct Threats adopted by most major international conservation organizations such as the US Nature Conservancy, the World Wildlife Fund, Conservation International, and Birdlife International.

**Destruction of Habitat**

Most of the species extinctions from 1000 AD to 2000 AD are due to human activities, in particular destruction of plant and animal habitats. Raised rates of extinction are being driven by human consumption of organic resources, especially related to tropical forest destruction. While most of the species that are becoming extinct are not food species, their biomass is converted into human food when their habitat is transformed into pasture, cropland, and orchards. It is estimated that more than a third of the Earth's biomass is tied up in only the few species that represent humans, livestock and crops.

Because an ecosystem decreases in stability as its species are made extinct, these studies warn that the global ecosystem is destined for collapse if it is further reduced in complexity. Factors contributing to loss of biodiversity are: overpopulation, deforestation, pollution (air pollution, water pollution, soil contamination) and global warming or climate change, driven by human activity. These factors, while all stemming from overpopulation, produce a cumulative impact upon biodiversity.

There are systematic relationships between the area of a habitat and the number of species it can support, with greater sensitivity to reduction in habitat area for species of larger body size and for those living at lower latitudes or in forests or oceans. Some characterize loss of biodiversity not as ecosystem degradation but by conversion to trivial standardized ecosystems (e.g., monoculture following deforestation). In some countries lack of property rights or access regulation to biotic resources necessarily leads to biodiversity loss (degradation costs having to be supported by the community).

A September 14, 2007 study conducted by the National Science Foundation found that biodiversity and genetic diversity are dependent upon each other—that diversity within a species is necessary to maintain diversity among species, and vice versa. According to the lead researcher in the study, Dr. Richard Lankau, "If any one type is removed from the system, the cycle can break down, and the community becomes dominated by a single species."

At present, the most threathened ecosystems are those found in fresh water. The marking of fresh water ecosystems as the ecosystems most under threat was done by the Millennium Ecosystem Assessment

2005, and was confirmed again by the project "Freshwater Animal Diversity Assessment", organised by the biodiversity platform, and the French Institut de recherche pour le development.

### Exotic Species

The rich diversity of unique species across many parts of the world exist only because they are separated by barriers, particularly large rivers, seas, oceans, mountains and deserts from other species of other land masses, particularly the highly fecund, ultra-competitive, generalist "super-species". These are barriers that couldn't have been easily crossed by natural processes, except through continental drift. However, humans have invented transportation with the ability to bring into contact species that they've never met in their evolutionary history; also, this is done on a time scale of days, unlike the centuries that historically have accompanied major animal migrations.

The widespread introduction of exotic species by humans is a potent threat to biodiversity. When exotic species are introduced to ecosystems and establish self-sustaining populations, the endemic species in that ecosystem that have not evolved to cope with the exotic species may not survive. The exotic organisms may be either predators, parasites, or simply aggressive species that deprive indigenous species of nutrients, water and light. These invasive species often have features, due to their evolutionary background and new environment, that make them highly competitive; able to become well-established and spread quickly, reducing the effective habitat of endemic species.

As a consequence of the above, if humans continue to combine species from different ecoregions, there is the potential that the world's ecosystems will end up dominated by relatively a few, aggressive, cosmopolitan "super-species". In 2004, an international team of scientists estimated that 10 percent of species would become extinct by 2050 because of global warming. "We need to limit climate change or we wind up with a lot of species in trouble, possibly extinct," said Dr. Lee Hannah, a co-author of the paper and chief climate change biologist at the Centre for Applied Biodiversity Science at Conservation International.

### Genetic Pollution

Purebred naturally evolved region specific wild species can be threatened with extinction through the process of genetic pollution i.e. uncontrolled hybridization, introgression and genetic swamping which leads to homogenization or replacement of local genotypes as a result of either a numerical and/or fitness advantage of introduced plant or animal. Nonnative species can bring about a form of extinction of native plants and animals by hybridization and introgression either through purposeful introduction by humans or through habitat modification, bringing previously isolated species into contact.

These phenomena can be especially detrimental for rare species coming into contact with more abundant ones. The abundant species can interbreed with the rarer, swamping the entire gene pool and creating hybrids, thus driving the entire native stock to complete extinction. Attention has to be focused on the extent of this under appreciated problem that is not always apparent from morphological (outward appearance) observations alone. Some degree of gene flow may be a normal, evolutionarily constructive, process, and all constellations of genes and genotypes cannot be preserved. However, hybridization with or without introgression may, nevertheless, threaten a rare species' existence.

## Hybridization and Genetics

In agriculture and animal husbandry, the green revolution popularized the use of conventional hybridization to increase yield by creating "high-yielding varieties". Often the handful of hybridized breeds originated in developed countries and were further hybridized with local varieties in the rest of the developing world to create high yield strains resistant to local climate and diseases. Local governments and industry have been pushing hybridization which has resulted in several of the indigenous breeds becoming extinct or threatened.

Disuse because of unprofitability and uncontrolled intentional and unintentional cross-pollination and crossbreeding (genetic pollution), formerly huge gene pools of various wild and indigenous breeds have collapsed causing widespread genetic erosion and genetic pollution. This has resulted in loss of genetic diversity and biodiversity as a whole.

A genetically modified organism (GMO) is an organism whose genetic material has been altered using the genetic engineering techniques generally known as recombinant DNA technology. Genetically Modified (GM) crops today have become a common source for genetic pollution, not only of wild varieties but also of other domesticated varieties derived from relatively natural hybridization.

Genetic erosion coupled with genetic pollution may be destroying unique genotypes, thereby creating a hidden crisis which could result in a severe threat to our food security. Diverse genetic material could cease to exist which would impact our ability to further hybridize food crops and livestock against more resistant diseases and climatic changes.

**Climate Change**

The recent phenomenon of global warming is also considered to be a major threat to global biodiversity. For example coral reefs -which are biodiversity hotspots- will be lost in 20 to 40 years if global warming continues at the current trend.

## CONSERVING BIODIVERSITY

A schematic image illustrating the relationship between biodiversity, ecosystem services, human well-being, and poverty. The illustration shows where conservation action, strategies and plans can influence the drivers of the current biodiversity crisis at local, regional, to global scales.

Conservation biology matured in the mid- 20th century as ecologists, naturalists, and other scientists began to collectively research and address issues pertaining to global declines in biodiversity. The conservation ethic differs from the preservationist ethic, historically lead by John Muir, who advocate for protected areas devoid of human exploitation or interference for profit.

The conservation ethic advocates for wise stewardship and management of natural resource production for the purpose of protecting and sustaining biodiversity in species, ecosystems, the evolutionary process, and human culture and society.

Conservation biologists are concerned with the trends in biodiversity being reported in this era, which has been labeled by science as the Holocene extinction period, also known as the sixth mass extinction. Rates of decline in biodiversity in this sixth mass extinction exceeds the five previous extinction spasms recorded in the fossil record. In response to the extinction crisis, the research of conservation biologists is being organized into strategic plans that include principles, guidelines, and tools for the purpose of protecting biodiversity.

Conservation biology is a crisis orientated discipline and it is multi-disciplinary, including ecological, social, education, and other scientific disciplines outside of biology. Conservation biologists work in both the field and office, in government, universities, non-profit organizations and in industry. The conservation of biological diversity is a global priority in strategic conservation plans that are designed to engage public policy and concerns affecting local, regional and global scales of communities, ecosystems, and cultures.

Conserving biodiversity and action plans identify ways of sustaining human well-being and global economics, including natural capital, market capital, and ecosystem services. One of the strategies involves placing a monetary value on biodiversity through biodiversity banking, of which one example is the Australian Native Vegetation Management Framework.

## JUDICIAL STATUS

Biodiversity is beginning to be evaluated and its evolution analysed (through observations, inventories, conservation...) as well as being taken into account in political and judicial decisions:

- The relationship between law and ecosystems is very ancient and has consequences for biodiversity. It is related to property rights, both private and public. It can define protection for threatened ecosystems, but also some rights and duties (for example, fishing rights, hunting rights).
- Law regarding species is a more recent issue. It defines species that must be protected because they may be threatened by extinction. The U.S. Endangered Species Act is an example of an attempt to address the "law and species" issue.
- Laws regarding gene pools are only about a century old. While the genetic approach is not new (domestication, plant traditional selection methods), progress made in the genetic field in the past 20 years have led to a tightening of laws in this field. With the new technologies of genetic analysis and genetic engineering, people are going through gene patenting, processes patenting, and a totally new concept of genetic resources. A very hot debate today seeks to define whether the resource is the gene, the organism itself, or its DNA.

The 1972 UNESCO World Heritage convention established that biological resources, such as plants, were the common heritage of mankind. These rules probably inspired the creation of great public banks of genetic resources, located outside the source-countries.

New global agreements (e.g. Convention on Biological Diversity), now give sovereign national rights over biological resources (not property). The idea of static conservation of biodiversity is disappearing and being replaced by the idea of dynamic conservation, through the notion of resource and innovation.

The new agreements commit countries to conserve biodiversity, develop resources for sustainability and share the benefits resulting from their use. Under new rules, it is expected that bioprospecting or collection of natural products has to be allowed by the biodiversity-rich country, in exchange for a share of the benefits.

Sovereignty principles can rely upon what is better known as Access and Benefit Sharing Agreements (ABAs). The Convention on Biodiversity spirit implies a prior informed consent between the source country and the collector, to establish which resource will be used and for what, and to settle on a fair agreement on benefit sharing. Bioprospecting can become a type of biopiracy when those principles are not respected.

Uniform approval for use of biodiversity as a legal standard has not been achieved, however. At least one legal commentator has argued that biodiversity should not be used as a legal standard, arguing that the multiple layers of scientific uncertainty inherent in the concept of biodiversity will cause administrative waste and increase litigation without promoting preservation goals.

## ANALYTICAL LIMITS

### Taxonomic and Size Bias

Less than 1% of all species that have been described have been studied beyond simply noting its existence. Biodiversity researcher Sean Nee points out that the vast majority of Earth's biodiversity is microbial, and that contemporary biodiversity physics is "firmly fixated on the visible world". For example, microbial life is very much more metabolically and environmentally diverse than multicellular life. Nee has stated: "On the tree of life, based on analyses of small-subunit ribosomal RNA, visible life consists of barely noticeable twigs.

The size bias is not restricted to consideration of microbes. Entomologist Nigel Stork states that "to a first approximation, all multicellular species on Earth are insects".

Even in insects, however, the extinction rate is high and indicative of the general trend of the sixth greatest extinction period that human society is faced with. Moreover, there are species co-extinctions, such as plants and beetles, where the extinction or decline in one is reciprocated in the other.

- Biodiversity is the variety of life: the different plants, animals and micro-organisms, their genes and the ecosystems of which they are a part. It is home to more than one million species of plants and animals, many of which are found nowhere else in the world.
- "Biodiversity" is often defined as the variety of all forms of life, from genes to species, through to the broad scale of ecosystems. "

## Chapter 2

# Conservation Biology

Conservation biology is the scientific study of the nature and status of Earth's biodiversity with the aim of protecting species, their habitats, and ecosystems from excessive rates of extinction. It is an interdisciplinary subject drawing on sciences, economics, and the practice of natural resource management. The term *conservation biology* was introduced as the title of a conference held University of California at San Diego in La Jolla, California in 1978 organized by biologists Bruce Wilcox and Michael Soulé.

The meeting was prompted by the growing concern among scientists over tropical deforestation, disappearing species, eroding genetic diversity within species. The conference and proceedings that resulted sought to bridge a gap existing at the time between theory in ecology and population biology on the one hand and conservation policy and practice on the other. Conservation biology and the concept of biological diversity (biodiversity) emerged together, helping crystallize the modern era of conservation science and policy.

The rapid decline of biological systems around the world means that conservation biology is often referred to as a "Discipline with a deadline". Conservation biology is tied closely to ecology in researching the dispersal, migration, demographics, effective population size, inbreeding depression, and minimum population viability of rare or endangered species.

Conservation biology is concerned with phenomena that affect the maintenance, loss, and restoration of biodiversity and the science of sustaining evolutionary processes that engender genetic, population, species, and ecosystem diversity. The concern stems from estimates suggesting that up to 50% of all species on the planet will disappear within the next 50 years, which has contributed to poverty, starvation, and will reset the course of evolution on this planet.

Conservation biologists research and educate on the trends and process of biodiversity loss, species extinctions, and the negative affect this is having on our capabilities to sustain the well-being of human society. Conservation biologists work in the field and office, in government, universities, non-profit organizations and industry. They are funded to research, monitor, and catalog every angle of the earth and its relation to society.

The topics are diverse, because this is an interdisciplinary network with professional alliances in the biological as well as social sciences. Those dedicated to the cause and profession advocate for a global response to the current biodiversity crisis based on morals, ethics, and scientific reason. Organizations and citizens are responding to the biodiversity crisis through conservation action plans that direct research, monitoring, and education programs that engage concerns at local through global scales.

## CONTEXT AND TRENDS

Conservation biologists study trends and process from the paleontological past to the ecological present as they gain an understanding of the context related to species extinction. It is generally accepted that there have been five major global mass extinctions that register in the planets history. These include: the Ordovician (440 mya), Devonian (370 mya), Permian–Triassic (245 mya), Triassic–Jurassic (200 mya), and Cretaceous (65 mya) extinction spasms.

Within the last 10,000 years, human influence over the Earth's ecosystems has been so extensive that scientists have difficulty estimating the number of species lost; that is to say the rates of deforestation, reef destruction, wetland draining and other human acts are proceeding much faster than human assessment of species. The latest Living Planet Report by the World Wide Fund for Nature estimates that we have exceeded the bio-regenerative capacity of the planet, requiring 1.5 Earths to support the current demands placed on our natural resources.

### The Sixth Extinction

Conservation biologists are dealing with and have published evidence from all corners of the planet indicating that humanity is living the sixth and greatest planetary extinction event of all time. We are living an era of unprecedented numbers of species extinctions, also known as the Holocene extinction event. The global extinction rate is approximately 100,000 times higher than the natural background extinction rate. It is estimated that two-thirds of all mammal genera and one-half of all mammal species weighing at least 44 kilograms (97 lb) have gone extinct in the last 50,000 years.

The Global Amphibian Assessment reports that amphibians are

declining on a global scale faster than any other vertebrate group, with over 32% of all surviving species being threatened with extinction. The surviving populations are in continual decline in 43% of those that are threatened. Since the mid-1980s the actual rates of extinction have exceeded 211 times rates measured from the fossil record.

However, "The current amphibian extinction rate may range from 25,039 to 45,474 times the background extinction rate for amphibians." The global extinction trend occurs in every major vertebrate group that is being monitored. For example, 23% of all mammals and 12% of all birds are Red Listed by the International Union for Conservation of Nature (IUCN), meaning they too are threatened with extinction.

## Status of Oceans and Reefs

Global assessments of coral reefs of the world continue to report drastic and rapid rates of decline. By 2000, 27% of the world's coral reef ecosystems had effectively collapsed. The largest period of decline occurred in a dramatic "bleaching" event in 1998, where approximately 16% of all the coral reefs in the world disappeared in less than a year. *Coral bleaching* is caused by a mixture of environmental stresses, including increases in ocean temperatures and acidity, causing both the release of symbiotic algae and death of corals. Conservation biology plays an important role in international agreements covering the world's oceans (and other issues pertaining to biodiversity).

The oceans are threatened most greatly by acidification due to an increase in $CO_2$ levels. This is a most serious threat to societies relying heavily upon oceanic natural resources. A concern is that the majority of all marine species will not be able to evolve or acclimate in response to the changes in the ocean chemistry.

The prospects of averting mass extinction seems unlikely when " 90% of all of the large (average approximately e"50 kg), open ocean tuna, billfishes, and sharks in the ocean" are reportedly gone. Given the scientific review of current trends, the ocean is predicted to have few surviving multicellular organisms with only microbes left to dominate marine ecosystems.

## Insects and Other Groups

There are serious concerns also being hailed from taxonomic groups that do not receive the same degree of social attention or attract funds as the vertebrates do, including fungi, lichen, plant and insect communities where the vast majority of biodiversity is represented. Insect conservation, in particular, is of pivotal importance for conservation biology.

The value of insects in the biosphere is enormous because they outnumber all other living groups in measure of species richness. The greatest bulk of biomass on land is found in plants, which is sustained by insect relations. This great ecological value of insects is countered by a society that often times reacts negatively toward these aesthetically 'unpleasant' creatures.

One area of concern in the insect world that has caught the public eye is the mysterious case of missing honey bees (*Apis mellifera*). Honey bees provide an indispensable ecological services through their acts of pollination supporting a huge variety of agriculture crops. The sudden disappearance of bees leaving empty hives or colony collapse disorder (CCD) is not uncommon. However, in a 16 month period from 2006 through 2007, 29% of 577 beekeepers across the United States reported CCD losses in up to 76% of their colonies. This sudden demographic loss in bee numbers is placing a strain on the agricultural sector. The cause behind the massive declines is puzzling scientists. Pests, pesticides, and global warming are all being considered as possible causes.

Another highlight that links conservation biology to insects, forests, and climate change is the mountain pine beetle (*Dendroctonus ponderosae*) epidemic of British Columbia, Canada, which has infested 470,000 km (180,000 sq mi) of forested land since 1999. An action plan has been prepared by the Government of British Columbia to address this problem.

This impact [*pine beetle epidemic*] converted the forest from a small net carbon sink to a large net carbon source both during and immediately after the outbreak. In the worst year, the impacts resulting from the beetle outbreak in British Columbia were equivalent to 75% of the average annual direct forest fire emissions from all of Canada during 1959–1999.

## Threats to Biodiversity

Many of the threats to biodiversity, including disease and climate change, are reaching inside borders of protected areas, leaving them 'not-so protected' (e.g. Yellowstone National Park). Climate change, for example, is often cited as a serious threat in this regard, because there is a feedback loop between species extinction and the release of carbon dioxide into the atmosphere.

Ecosystems store and cycle large amounts of carbon to regulate global conditions. The effects of global warming adds a catastrophic threat toward a mass extinction of global biological diversity. The extinction threat is estimated to range from 15 to 37 percent of all species by 2050, or 50 percent of all species over the next 50 years. Some of the most significant and insidious threats to biodiversity and ecosystem processes include climate change, mass agriculture, deforestation, overgrazing, slash-and-burn agriculture, urban development, wildlife trade, light pollution and pesticide use. Habitat fragmentation poses one of the more difficult challenges, because the global network of protected areas only covers 11.5% of the Earth's surface.

A significant consequence of fragmentation and lack of linked protected areas is the reduction of animal migration on a global scale. Considering that billions of tonnes of biomass are responsible for nutrient cycling across the earth, the reduction of migration is a serious matter for conservation biology.

These figures do not imply, however, that human activities must necessarily cause irreparable harm to the biosphere. With conservation management and planning for biodiversity at all levels, from genes to ecosystems, there are examples where humans mutually coexist in a sustainable way with nature. However, it may be too late for human intervention to reverse the current mass extinction.

## Concepts and Foundations

### *Measuring Extinction Rates*

Extinction rates are measured in a variety of ways. Conservation biologists measure and apply statistical measures of fossil records, rates of habitat loss, and a multitude of other variables such as loss of biodiversity as a function of the rate of habitat loss and site occupancy to obtain such estimates. The Theory of Island Biogeography is possibly the most significant contribution toward the scientific understanding of both the process and how to measure the rate of species extinction. The current background extinction rate is estimated to be one species every few years.

The measure of ongoing species loss is made more complex by the fact that most of the Earth's species have not been described or evaluated.

Estimates vary greatly on how many species actually exist (estimated range: 3,600,000-111,700,000) to how many have received a species binomial (estimated range: 1.5-8 million). Less than 1% of all species that have been described have been studied beyond simply noting its existence. From these figures, the IUCN reports that 23% of vertebrates, 5% of invertebrates and 70% of plants that have been evaluated are designated as endangered or threatened.

## Systematic Conservation Planning

Systematic conservation planning is an effective way to seek and identify efficient and effective types of reserve design to capture or sustain the highest priority biodiversity values and to work with communities in support of local ecosystems. Margules and Pressey identify six interlinked stages in the systematic planning approach:

- Compile data on the biodiversity of the planning region.
- Identify conservation goals for the planning region.
- Review existing conservation areas.
- Select additional conservation areas.
- Implement conservation actions.
- Maintain the required values of conservation areas.

Conservation biologists regularly prepare detailed conservation plans for grant proposals or to effectively coordinate their plan of action and to identify best management practices (e.g.). Systematic strategies generally employ the services of Geographic Information Systems to assist in the decision making process.

## Conservation Biology as a Profession

The Society for Conservation Biology is a global community of conservation professionals dedicated to advancing the science and practice of conserving biodiversity. Conservation biology as a discipline reaches beyond biology, into subjects such as philosophy, law, economics, humanities, arts, anthropology, and education. Within biology, conservation genetics and evolution are immense fields unto themselves, but these disciplines are of prime importance to the practice and profession of conservation biology.

Is conservation biology an objective science when biologists advocate for an inherent value in nature? Do conservationists introduce bias when they support policies using qualitative description, such as habitat *degradation,* or *healthy* ecosystems? As all scientists hold values, so do conservation biologists. Conservation biologists advocate for reasoned and sensible management of natural resources and do so with a disclosed combination of science, reason, logic, and values in their conservation management plans.

This sort of advocacy is similar to the medical profession advocating for healthy lifestyle options, both are beneficial to human well-being yet remain scientific in their approach. Many conservation biologists, in addition to having a Bachelors of Science (or extensive natural experience) often receive professional accreditation during their career.

There is a movement in conservation biology suggesting a new form of leadership is needed to mobilize conservation biology into a more effective discipline that is able to communicate the full scope of the problem to society at large. The movement proposes an adaptive leadership approach that parallels an adaptive management approach. The concept is based on a new philosophy or leadership theory steering away from historical notions of power, authority, and dominance.

Adaptive conservation leadership is reflective and more equitable as it applies to any member of society who can mobilize others toward meaningful change using communication techniques that are inspiring, purposeful, and collegial. Adaptive conservation leadership and mentoring programs are being implemented by conservation biologists through organizations such as the Aldo Leopold Leadership Programme

## Preservation

A preservationist differs from a conservation biologist through the principal of non-interference. Preservationists advocate for giving areas of nature and species a protected existence that halts interference from the humans. In this regard, conservationists differ from preservationists in the social dimension, as conservation biology engages society and seeks equitable solutions for both society and ecosystems.

## Ethics and Values

Conservation biologists are interdisciplinary researchers that practice ethics in the biological and social sciences. Chan states that conservationists must advocate for biodiversity and can do so in a scientifically ethical manner by not promoting simultaneous advocacy against other competing values. A conservationist researches biodiversity and reasons through a Resource Conservation Ethic, which identify what measures will deliver "the greatest good for the greatest number of people for the longest time."

Some conservation biologists argue that nature has an intrinsic value that is independent of anthropocentric usefulness or utilitarianism. Intrinsic value advocates that a gene, or species, be valued because they have a utility for the ecosystems they sustain. Aldo Leopold was a classical thinker and writer on such conservation ethics whose philosophy, ethics and writings are still valued and revisited by modern conservation biologists. His writing is oftentimes required reading for those in the profession.

## Conservation Priorities

While most in the community of conservation science "stress the importance" of sustaining biodiversity, there is debate on how to prioritize genes, species, or ecosystems, which are all components of biodiversity.

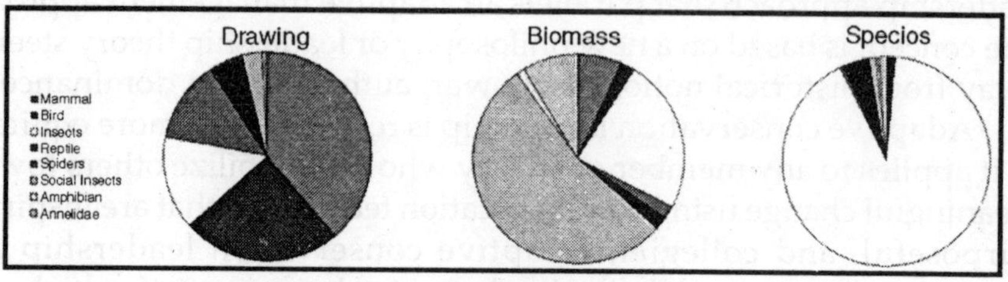


**Fig.** A Pie Chart Image Showing the Relative Biomass Representation in a Rain Forest Through a Summary of Children's Perceptions from Drawings and Artwork (left), Through a Scientific Estimate of Actual Biomass (middle), and by a Measure of Biodiversity (right). Notice that the Biomass of Social Insects (middle) far Outweighs the Number of Species (right).

While the predominant approach to date has been to conserve *biodiversity hotspots*, some scientists (e.g.) and conservation organizations, such as the Nature Conservancy, argue that it is more cost effective and socially relevant to invest in *biodiversity coldspots*.

Biodiversity hotspots and coldspots are a way of recognizing that the spatial concentration of genes, species, and ecosystems is not uniformly distributed on the earth's surface. For example, " 44% of all species of vascular plants and 35% of all species in four vertebrate groups are confined to 25 hotspots comprising only 1.4% of the land surface of the Earth."

Those arguing in favour of setting priorities for coldspots point out that there are other measures to consider beyond biodiversity. They point out that emphasizing hotspots downplays the importance of the social and ecological connections to vast areas of the Earth's ecosystems where biomass, not biodiversity, reigns supreme. It is estimated that 36% of the Earth's surface, encompassing 38.9% of the worlds vertebrates, lacks the endemic species to qualify as biodiversity hotspot.

Moreover, measures show that maximizing protections for biodiversity does not capture ecosystem services any better than targeting randomly chosen regions. Population level biodiversity (i.e. coldspots) are disappearing at a rate that is ten times that at the species level. The level of importance in addressing biomass versus endemism as a concern for conservation biology is highlighted in literature measuring the level of threat to global ecosystem carbon stocks that do not necessarily reside in areas of endemism.

A hotspot priority approach would not invest so heavily in places such as steppes, the Serengeti, the Arctic, or taiga. These areas contribute a great abundance of population (not species) level biodiversity and ecosystem services, including cultural value and planetary nutrient cycling.

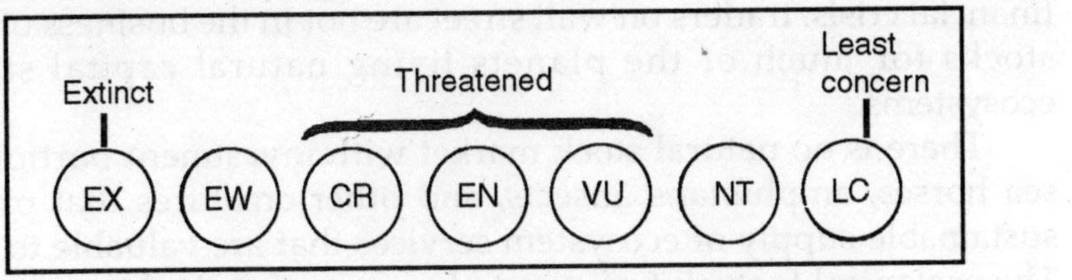


**Fig.** Summary of 2006 IUCN Red List Categories.

Those in favour of the hotspot approach point out that species are irreplaceable components of the global ecosystem, they are concentrated in places that are most threatened, and should therefore receive maximal strategic protections. The IUCN Red List categories, which appear on Wikipedia species articles, is an example of the hotspot conservation approach in action; species that are not rare or endemic are listed the least concern and their wikipedia articles tend to be ranked low on the importance scale.

This is a hotspot approach because the priority is set to target species level concerns over population level or biomass. Species richness and genetic biodiversity contributes to and engenders ecosystem stability, ecosystem processes, evolutionary adaptability, and biomass. Both sides agree, however, that conserving biodiversity is necessary to reduce the extinction rate and identify an inherent value in nature; the debate hinges on how to prioritize limited conservation resources in the most cost effective way.

## Economic Values and Natural Capital

Conservation biologists have started to collaborate with leading global economists to determine how to measure the wealth and services of nature and to make these values apparent in global market transactions. This system of accounting is called *natural capital* and would, for example, register the value of an ecosystem before it is cleared to make way for development.

The WWF publishes its *Living Planet Report* and provides a global index of biodiversity by monitoring approximately 5000 populations in 1,686 species of vertebrate (mammals, birds, fish, reptiles, and amphibians) and report on the trends in much the same way that the stock market is tracked.

This method of measuring the global economic benefit of nature has been endorsed by the G8+5 leaders and the European Commission. Nature sustains many ecosystem services that benefit humanity. Many of the earths ecosystem services are public goods without a market and therefore no price or value. When the *stock market* registers a financial crisis, traders on wall street are not in the business of trading stocks for much of the planets living natural capital stored in ecosystems.

There is no natural stock market with investment portfolios into sea horses, amphibians, insects, and other creatures that provide a sustainable supply of ecosystem services that are valuable to society. The ecological footprint of society has exceeded the bio-regenerative capacity limits of the planets ecosystems by about 30 percent, which is the same percentage of vertebrate populations that have registered decline from 1970 through 2005.

The inherent natural economy plays an essential role in sustaining humanity, including the regulation of global atmospheric chemistry, pollinating crops, pest control, cycling soil nutrients, purifying our water supply, supplying medicines and health benefits, and unquantifiable quality of life improvements.

There is a relationship, a correlation, between markets and natural capital, and social income inequity and biodiversity loss. This means that there are greater rates of biodiversity loss in places where the inequity of wealth is greatest

Although a direct market comparison of natural capital is likely insufficient in terms of human value, one measure of ecosystem services suggests the contribution amounts to trillions of dollars yearly. For example, one segment of North American forests has been assigned an annual value of 250 billion dollars; as another example, honey-bee pollination is estimated to provide between 10 and 18 billion dollars of value yearly.

The value of ecosystem services on one New Zealand island has been imputed to be as great as the GDP of that region. This planetary wealth is being lost at an incredible rate as the demands of human society is exceeding the bio-regenerative capacity of the Earth. While biodiversity and ecosystems are resilient, the danger of loosing them is that humans cannot recreate many ecosystem functions through technological innovation.

## STRATEGIC SPECIES CONCEPTS

### Keystone Species

Some species, called a *keystone species*, form a central supporting hub in the ecosystem. The loss of such a species results in a collapse in ecosystem function, as well as the loss of coexisting species. The importance of a keystone species was shown by the extinction of the Steller's Sea Cow (*Hydrodamalis gigas*) through its interaction with sea otters, sea urchins, and kelp. Kelp beds grow and form nurseries in shallow waters to shelter creatures that support the food chain.

Sea urchins feed on kelp, while sea otters feed on sea urchins. With the rapid decline of sea otters due to overhunting, sea urchin populations grazed unrestricted on the kelp beds and the ecosystem collapsed. Left unchecked, the urchins destroyed the shallow water kelp communities that supported the Steller's Sea Cow's diet and hastened their demise. The sea otter is a keystone species because the coexistence of many ecological associates in the kelp beds relied upon otters for their survival.

## Indicator Species

An *indicator species* has a narrow set of ecological requirements, therefore they become useful targets for observing the health of an ecosystem. Some animals, such as amphibians with their semi-permeable skin and linkages to wetlands, have an acute sensitivity to environmental harm and thus may serve as a *miner's canary*. Indicator species are monitored in an effort to capture environmental degradation through pollution or some other link to proximate human activities.

Monitoring an indicator species is a measure to determine if there is a significant environmental impact that can serve to advise or modify practice, such as through different forest silviculture treatments and management scenarios, or to measure the degree of harm that a pesticide may impart on the health of an ecosystem.

Government regulators, consultants, or NGOs regularly regularly monitor indicator species, however, there are limitations coupled with many practical considerations that must be followed for the approach to be effective. It is generally recommended that multiple indicators (genes, populations, species, communities, and landscape) be monitored for effective conservation measurement that prevents harm to the complex, and oftentimes unpredictable, response from ecosystem dynamics.

## Umbrella and Flagship Species

An example of an *umbrella species* is the Monarch butterfly, because of its lengthy migrations and aesthetic value. The Monarch migrates across North America, covering multiple ecosystems and so requires a large area to exist. Any protections afforded to the Monarch butterfly will at the same time umbrella many other species and habitats. An umbrella species is often used as *flagship species*, which are species, such as the Giant Panda, Blue Whale, tigers, and the Monarch butterfly, that capture the public's attention and attract support for conservation measures.

## Natural Resource Conservation

Efforts to conserve and protect *global* biodiversity is a recent phenomenon. Prior to the global conservation era, there was the coming of the age of conservation. Some historians have linked with the 1916 National Parks Act, which included the 'use without impairment' clause, sought by John Muir. This eventually resulted in the removal of a proposal to build a dam in Dinosaur National Monument in 1959.

Natural resource conservation, however, has a history that extends prior to the age of conservation. Resource ethics grew out of necessity through direct relations with nature. Regulation or communal restraint became necessary to prevent selfish motives from taking more than could be locally sustained, therefore compromising the long-term supply for the rest of the community. This social dilemma with respect to natural resource management is often called the "Tragedy of the Commons."

From this principal, conservation biologists can trace communal resource based ethics throughout cultures as a solution to communal resource conflict. For example, the Alaskan Tlingit peoples and the Haida of the Pacific Northwest had resource boundaries, rules, and restrictions among clans with respect to the fishing of Sockeye Salmon. These rules were guided by clan elders who knew life-long details of each river and stream they managed. There are numerous examples in history where cultures have followed rules, rituals, and organized practice with respect to communal natural resource management.

Conservation ethics are also found in early religious and philosophical writings. There are examples in the Tao, Shinto, Hindu, Islamic and Buddhist traditions. In Greek philosophy, Plato lamented about pasture land degradation: "What is left now is, so to say, the skeleton of a body wasted by disease; the rich, soft soil has been carried off and only the bare framework of the district left." In the bible, through Moses, God commanded to let the land rest from cultivation every seventh year.

Before the 18$^{th}$ century, however, much of European culture considered it a pagan view to admire nature. Wilderness was denigrated while agricultural development was praised. Even in 18th century America, However, as early as AD 680 a wildlife sanctuary was founded on the Farne Islands by St Cuthbert in response to his religious beliefs.

**Early Naturalists**

Natural history was a major preoccupation in the 18th century, with grand expeditions and the opening of popular public displays in Europe and North America. By 1900 there were 150 natural history museums in Germany, 250 in Great Britain, 250 in the United States, and 300 in France. Preservationist or conservationist sentiments are a development in the late 18th to early 20th century. The 19th century fascination with natural history engendered a fervor to be the first to collect rare specimens with the goal of doing so before they became extinct by other such collectors.

Although his artistic works and romantic depiction of avian life inspired many bird enthusiasts and conservation organizations, the writings of John James Audubon, by modern standards, show insensitivity toward bird conservation as he shot and collected hundreds of specimens. Inspired by him, however, the first chapter of the Audubon Society started in 1905 for the purpose of protecting birds.

**Coming of the Age of Conservation**

The modern concept of ecosystem services can be found in the late 19th century. "The utility of Natural History or its applicability to promote the material wealth of the State cannot be doubted. It was a great mistake to suppose that the subjects of Zoology, Botany, and Geology did not involve much that affects our comfort, convenience, health and wealth." However, the article continues and discusses the dread of agricultural pests and the utility of understanding their natural history for the purpose of facilitating their destruction.

By the early 1800s biogeography was ignited through efforts of Von Humboldt, DeCandolle, Lyell and Darwin; their efforts, while important in relating species to their environments, were part of the naturalist tradition and fell short of conservation biology proper. Darwin, for example, hunted and shot birds and kept natural history cabinets in line with Victorian tradition.

Modern roots of conservation biology can be found in the late 19th century Enlightenment period particularly in England and Scotland. A number of thinkers, among them notably Lord Monboddo, described the importance of "preserving nature"; much of this early emphasis had its origins in Christian theology.

## 20th Century Conservation

In the 20th century, actions in the United Kingdom, United States, and Canada emphasized the protection of habitat areas pursuant to visions of such people as John Muir, Theodore Roosevelt, and Aldo Leopold. While the Canadian nor the United Kingdom governments did not pioneer the creation of National Parks as the United States did in the late 19th century, there were many far-sighted civil servants who were dedicated to wildlife conservation and of notable mention. Some of these historical figures include Charles Gordon Hewitt and James Harkin.

The term *conservation* came into use in the late 19th century and referred to the management, mainly for economic reasons, of such natural resources as timber, fish, game, topsoil, pastureland, and minerals. In addition it referred to the preservation of forests (forestry), wildlife (wildlife refuge), parkland, wilderness, and watersheds. Western Europe was the source of much 19th century progress for conservation biology, particularly the British Empire with the Sea Birds Preservation Act 1869. However, the United States made contributions to this field starting with thinking of Thoreau and taking form with the Forest Act of 1891, John Muir's founding of the Sierra Club in 1892, the founding of the New York Zoological Society in 1895 and establishment of a series of national forests and preserves by Theodore Roosevelt from 1901 to 1909.

Not until the mid 20th century did efforts arise to target individual species for conservation, notably efforts in big cat conservation in South America led by the New York Zoological Society. In the early 20th century the New York Zoological Society was instrumental in developing concepts of establishing preserves for particular species and conducting the necessary conservation studies to determine the suitability of locations that are most appropriate as conservation priorities; the work of Henry Fairfield Osborn Jr., Archie Carr and Archie Carr III is notable in this era.

By the 1970s, led primarily by work in the United States under the Endangered Species Act along with the Species at Risk Act (SARA) of Canada, Biodiversity Action Plans developed in Australia, Sweden, the United Kingdom, hundreds of species specific protection plans ensued. Notably the United Nations acted to conserve sites of outstanding cultural or natural importance to the common heritage of mankind. The programme was adopted by the General Conference of UNESCO in 1972.

As of 2006, a total of 830 sites are listed: 644 cultural, 162 natural. The first country to pursue aggressive biological conservation through national legislation was the United States, which passed back to back legislation in the Endangered Species Act and National Environmental Policy Act, which together injected major funding and protection measures to large scale habitat protection and threatened species research. Other conservation developments, however, have taken hold throughout the world. India, for example, passed the Wildlife Protection Act of 1972.

In 1980 a significant development was the emergence of the urban conservation movement. A local organization was established in Birmingham, UK, a development followed in rapid succession in cities across the UK, then overseas. Although perceived as a grassroots movement, its early development was driven by academic research into urban wildlife.

Initially perceived as radical, the movement's view of conservation being inextricably linked with other human activity has now become mainstream in conservation thought. Considerable research effort is now directed at urban conservation biology. The Society for Conservation Biology originated in 1985.

By 1992 most of the countries of the world had become committed to the principles of conservation of biological diversity with the Convention on Biological Diversity; subsequently many countries began programmes of Biodiversity Action Plans to identify and conserve threatened species within their borders, as well as protect associated habitats. The late 1990s saw increasing professionalism in the sector, with the maturing of organisations such as the Institute of Ecology and Environmental Management and the Society for the Environment.

Since 2000 the concept of landscape scale conservation has risen to prominence, with less emphasis being given to single-species or even single-habitat focused actions. Instead an ecosystem approach is advocated by most mainstream conservationist, although concerns have been expressed by those working to protect some high-profile species.

Ecology has clarified the workings of the biosphere; i.e., the complex interrelationships among humans, other species, and the physical environment. The burgeoning human population and associated agriculture, industry, and the ensuing pollution, have demonstrated how easily ecological relationships can be disrupted.

Chapter 3

# Biodiversity and International Water Policy

## DEVELOPMENT OF GLOBAL FRESHWATER BIODIVERSITY

In the past, human-induced pressures on freshwater ecosystems were rarely significant and mostly of a local nature. In general, the resilience of freshwater ecosystems was sufficient to recover from such stress caused by human intrusion. But this situation has fundamentally changed in the last decades. In the past fifty years, for example, rapid population growth in conjunction with technological advances, the expansion of irrigated land, the spread of high-input agriculture, deforestation, modified land-use pattern, and changing consumer behaviour have had an ever-greater impact on freshwater ecosystems. Other human activities disrupt the dynamic hydrological regime of ecosystems.

Dams, for example, directly and indirectly influence the ecological integrity of downstream freshwater ecosystems and might contribute significantly to the loss of biodiversity in certain areas. Further, human-induced emissions of greenhouse gases have apparently contributed to global climate change affecting freshwater ecosystem integrity in many regions of the world. In general, human-induced impacts on water quantity, quality and flow have caused environmental degradation, destruction of natural habitats and loss of their ecological functions. Indirect landscape and direct waterscape changes have had important impacts including:

- Levelling of seasonal flow patterns and reduction of river flow volume.
- Blockage of access of species to river segments.
- Input of organic and - toxic matter.
- Increased nutrient levels.
- Turbidity and sediments.

However, water experts are used to tackling water pollution and supply and have been (partly) successful in doing so. Today, in the developed world pollution from industrial waste and domestic sewage is decreasing while pollution from non-point sources is rather on the increase. In general, in Europe and North America, pollution, acidification, and the physical modification of streams have had the greatest impact on the biodiversity of freshwater ecosystems. In the developing world, collection and adequate treatment of domestic sewages, and industrial pollution still is a major problem, especially in rapidly growing urban areas.

Whereas the public awareness regarding the pollution of freshwater resources has grown, other factors threatening freshwater ecosystems, especially those relating to the interaction of water and land-use management or the impacts of non-native species, have received much less attention.

While the state of global freshwater biodiversity is far from being fully explored from a scientific point of view, there are clear indications that freshwater ecosystems experience a significant loss of their biological diversity due to various human activities. Available data suggest that around 30% of the fish species are threatened, mostly because of habitat destruction.

The introduction of alien species, pollution, and over exploitation – between 1961 and 1996 there was a five-fold increase in the worldwide freshwater fish catch – are also important factors. In Europe 42% of freshwater fish are threatened, in Iran 22%, and in South Africa 63%.

The proportion of threatened species of freshwater mammals is about 65%, those of the world's amphibians approximately 25%. Equally, many birds and mammals depending on freshwater have lost their main breeding and feeding areas due to the alteration of freshwater ecosystems. Furthermore, according to recent estimations earth has lost about 50% of its original wetlands in the last century.

In general, it has been shown that freshwater biodiversity is more threatened than terrestrial ecosystems' biodiversity – even more than the much publicised plight of the world's forests. In recent years, attitudes towards the integration of nature protection aspects in water policy have changed as biodiversity loss of freshwater is increasingly perceived as a key challenge for water management practices in many countries.

This can be traced back at least to two reasons: First, the recognition that the ongoing alteration of "natural" freshwater ecosystems is not a sustainable option in the long term. Second, the insight that less modified freshwater ecosystems and the preservation of freshwater biodiversity provides significant economic and social benefits to society.

To mention one example, floods have been regarded by land-use planners and water experts as a waste and a threat to human economic activities. During the last two decades it is increasingly recognised that in many cases floods are highly productive in agricultural terms; ensure regular recharge of groundwater aquifers; have valuable ecological functions in terms of biodiversity protection and sometimes important pay-offs for recreation and tourism. In general, freshwater ecosystems provide a multitude of goods and services such as water supply, flow regulation, waste assimilation, and regulation of climate and the atmospheric chemical composition.

Further, freshwater ecosystems provide essential goods for human well-being, e.g. food, power, and raw material production as well as recreation. In spite of methodical difficulties in economic valuation, intrinsic values or non-use services such as cultural, aesthetic or scientific values are also very important, even if such non-use values might be more relevant in wealthy nations. In particular, biodiversity of ecosystems provide economically important services, especially in the long-term, because loss of diversity decreases the probability of natural recovery of freshwater ecosystems after human-induced or natural disturbances.

Therefore, in many cases the preservation of biodiversity will prevent future costs resulting from the degradation of water ecosystems and keep future economic development options open. Meanwhile there are many studies which attempt to quantify precisely the economic benefits resulting from the integrity of freshwater resources: For example, Barbier and Thompson have shown that the reduction of inundated areas in the Hadajia-Nguru wetlands (Nigeria) for intensive agriculture purposes led to net economic losses; the economic returns of an intensive irrigation scheme (Kano River project) had a significantly lesser volume than the calculated economic benefits of the floodplain, without even taking into account intrinsic values.

Sather and Smith have estimated the economic value of flood prevention at US$ 13,500 hectare/year provided by wetlands in the catchment of the Charles River in Massachusetts. New York City's investments in the purchase and management of land in its water catchment is a good example of cost-savings due to expansion of freshwater ecosystems for water purification purposes. In this case, New York City could not only avoid costs amounting US$ 6,5 billion but was also able to provide new wildlife habitats, other ecological resources and additional areas for recreation.

According to Costanza *et al.* global economic benefits of freshwater ecosystems is several times higher per unit area compared to terrestrial ecosystems as forests, grasslands or rangelands. Other scientists have estimated, that the global wetland systems alone are worth some US$ 4.0 trillion a year! However, though the recent academic and political consensus at the global level indicates that the economic and social benefits of the protection of freshwater ecosystems are increasingly taken into account, more efforts are needed to translate this consensus into actual water management practices.

In general, the present set of regulations and management tools in practice are considered inadequate. In addition, it is worth emphasising that priorities in water management in the developed and developing world are rather different. In the developing world the priorities are elementary improvements in health, food security and water supply. But there are clear indications that the satisfaction of those elementary needs presupposes the preservation of the integrity of freshwater ecosystems. Therefore, the perception of ecological aspects in water policy in some developing countries as a "northern" topic is at least over-accentuated because in many cases people in rural areas in particular suffer from freshwater ecosystem degradation.

**Table. Global Freshwater Biodiversity "Hot Spots": Freshwater Fish Species**

| Countries with most species | Number of species | Countries with mostspecies per km² | Number of species/1000 km² |
|---|---|---|---|
| Brazil | 3,000 | Burundi | 8.15 |
| Indonesia | 1,300 | Malawi | 3.84 |
| China | 1,010 | Bangladesh | 2.00 |
| Zaire | 962 | Malaysia | 1.83 |
| Peru | 855 | Sierra Leone | 1.63 |
| United States | 779 | Lao P.D.R. | 1.48 |
| India | 748 | Cambodia | 1.47 |
| Thailand | 690 | Vietnam | 1.38 |
| Tanzania | 682 | Thailand | 1.35 |
| Malaysia | 600 | Uganda | 1.24 |

Many river basins in the developing world with high biodiversity are vulnerable to future pressures (e.g. the Nile, the Volta, the Irrawaddy and the Mekong) and additional financial resources for developing countries seem to be necessary for the protection of the global biodiversity of freshwater resources. In the developed world, the emphasis is more on protecting the remaining natural and semi-natural freshwater ecosystems and on restoring degraded ecosystems (e.g. floodplains, wetlands) respectively.

Equally, although most natural freshwater ecosystems are highly modified, there are many hot spots of freshwater biodiversity in the developed world, such as the Mississippi Basin, and many other rivers and lakes containing a considerable number of endemic species. In general, despite the difference in priorities and institutional prerequisites in the developing and the developed world, the general challenge to increase the consideration of biodiversity aspects in water management practices applies to both.

## GLOBAL RESPONSES TO THE LOSS OF FRESHWATER BIODIVERSITY

The world community has recognised the loss of freshwater biodiversity in the last decade and has started initiatives to tackle the causes of environmental degradation of freshwater ecosystems. The three Rio global environmental conventions (biodiversity, climate, desertification), the Ramsar Convention on Wetlands, and the Convention on International Watercourses, have altogether set a new international framework to reverse the trend of ongoing environmental degradation.

### Increasing International Consensus on Management Principles

Water chapter, requires from the States "to make certain that adequate supplies of water of good quality are maintained for the entire population of this planet, while preserving the hydrological, biological and chemical functions of ecosystems, adapting human activities within the capacity limits of nature". The programme area "Protection of water resources, water quality and aquatic ecosystems" in particular mentions the many human and economic uses that disturb aquatic ecosystems and threaten living freshwater resources, e.g. agricultural water resource development projects such as dams, river diversion, water installations and irrigation schemes, and calls - with reference to the Mar del Plata Action Plan - for evaluating the consequences which the various users of water have on the environment, and to protect ecosystems.

Maintaining ecosystem integrity is but one objective of water resource management either in national or transboundary river systems "including living resources, and of effectively protecting them from any form of degradation on a drainage basin basis". The international community by adopting Agenda has set ambitious targets for the protection of ecosystems:

- To initiate programs for the protection and conservation of these resources on a sustainable basis.
- To initiate effective water pollution prevention and control programs based on an appropriate mix of pollution reduction-at-source strategies, environmental impact assessments and enforceable standards for major point-source discharges and high-risk non-point sources.
- To establish according to capacities and needs, biological, health, physical and chemical quality criteria for all water bodies.
- To adopt an integrated approach to environmentally sustainable management of water resources, including the protection of aquatic ecosystems and freshwater living resources.
- To put in place strategies for the environmentally sound management of freshwater and related coastal ecosystems, including consideration of fisheries, aquaculture, animal grazing, agricultural activities and biodiversity.

Agenda, water chapter, further recommends a bouquet of activities to protect freshwater living resources from pollution and degradation, to restore aquatic habitats and ecosystems, and to conserve and protect "wetlands (owing to their ecological and habitat importance for many species), taking into account social and economic factors; control of noxious aquatic species that may destroy some other water species". Clearly, Agenda, water chapter, is a declaration of intent and uses a more vague language concerning particular problems which can be traced back to different perceptions of developed and developing countries of the nature of the environmental problem.

Further progress was made through the adoption of the CSD-6 document "Strategic approaches to freshwater management" in 1998. Throughout this document, environmental effects of water management are more systematically considered and the linkages of water to biodiversity and land degradation are explicitly mentioned. Thus, while having adopted Agenda, water chapter, and subsequent international agreements, states have already found a more or less adequate language for biodiversity-related aspects of water management.

But due to its political character, one important shortcoming of the Agenda is the far-reaching neglect of biodiversity aspects in the management of transboundary water resources. Therefore, the water chapter provides a useful framework for reforms leading water management towards sustainability in general terms, but it does not constitute a regulatory framework for interaction of riparian states concerning preservation of freshwater biodiversity.

**Convention on International Watercourses**

The Convention on the Law of the Non-navigational Uses of International Watercourses is the basic document of international water law that guides riparian States for the peaceful regulation of competing demands such as flood control, water diversion for irrigation, generation of hydro power, pollution control, and for freshwater ecosystem protection. During the course of negotiation it has been argued that the Convention has not successfully addressed recent environmental challenges and would not take due consideration of protecting and preserving transboundary rivers' ecosystems, flora and fauna in adjacent areas, estuaries and wetlands.

Whether the Convention is conducive towards the protection of freshwater ecosystems can be judged from the physical scope of its applicability, the relation between the principles of equitable, reasonable utilisation and of no-harm and its environmental provisions.

***The "international Watercourse" Concept***

The geographical and hydrological area within which the Convention is operative is defined by the term "international watercourse", i.e. "a system of surface waters and ground waters constituting by virtue of their physical relationship a unitary whole and normally flowing into a common terminus; parts of which are situated in different States". In making an "international watercourse" the point of reference, the Convention has replaced the early "drainage basin" concept of the Helsinki Rules and that of a "shared natural resource" which was introduced at an early stage of negotiations. The "shared natural resource" concept would have meant that States do not have exclusive control over their portions of a river and would have required a certain degree of internationalisation.

The geographical implications of the term "drainage basin" too, led to abandon it, because States feared it could open the way to undue restrictions on their sovereignty over water resources and surrounding land areas. Only the "international watercourse" approach has found sufficient support from within the international community. However, the concept was broadened by the above mentioned definition extending a watercourse beyond the main surface water channel of a river, but still lacks mentioning the interconnectedness between e.g. land and water, and ecosystems concern is limited to the watercourse itself.

## Equitable, Reasonable Utilisation and the No-harm Principle

These Articles are the cornerstone of the Convention, and debates centred on whether the no-harm rule or the principle of equitable, reasonable utilisation should be superior. While, as a rule, upper riparian countries have favoured the principle of equitable and reasonable utilisation, downstream and environmentally concerned countries have emphasised that "equitable utilisation" should not serve to justify environmental damage.

However, even if the no-harm rule would have gained precedence, which it has not, the Convention does not prohibit harm in absolute terms, but rather relies on a due diligence obligation, and puts a ban on activities causing *significant* harm. While some transboundary impacts may not negatively affect downstream human and economic uses, it may impact negatively on ecosystems. A stricter ecosystem orientation requires setting e.g. emission standards, limit values or any other qualification pursuant to ecological rather than only human or economic factors.

## Environmental Provisions

Part IV provides for protecting and preserving the *ecosystems* of international watercourses. Although not defined in the Convention, the commentary considers that the term *ecosystem* refers to ecological units consisting of living and non-living components that are interdependent and function as a community. Consequently, the environmental provisions call for the preventing, reducing and controlling of pollution in cases that may cause significant harm to other watercourse States or their environment, including harm to human health or safety, to the use of waters for any beneficial purpose or to the living resources of the watercourse.

Watercourse States shall take steps to harmonise their policies in this connection. The Convention indicates an increased emphasis on co-operation which has led to the creation of a list of potential pollution abatement measures that States, at the request of one of them, could consult on. It recommends to set joint water quality objectives and criteria; to establish techniques and practices that address pollution from point and non-point sources; to establish a list of substances, the introduction of which into international watercourses is to be prohibited, limited, investigated and monitored.

More importantly, Article shows that the obligation for prevention, reducing and controlling the pollution of international watercourses does not depend on whether significant harm is caused to co-riparian states, but that protective measures may be necessary *within* the States' territories even if no pollution is caused to actual or potential uses of States.

A specific obligation of Article is that States are required to do all that can be reasonably expected to prevent the introduction of alien and new species as foreign organisms can upset the ecological balance resulting in the acceleration of eutrophication, elimination of other species, disruption of food webs, transmission of diseases, clogging of intakes etc. In addition, the Convention mentions the protection and preservation of the marine environment "since much of the most serious damage is caused by pollution from landbased sources, including rivers".

Article calls upon riparian States for "Prevention and Mitigation of Harmful Conditions" which requires that States take reasonable measures for controlling, among other things, deforestation to prevent downstream floods, siltation, erosion, salt water intrusion, drought or desertification.

Concerning planned measures, there is no obligation for an Environmental Impact Assessment (EIA), and states are only obliged to provide the result of any EIA if conducted; this also applies to the impact of planned measures on ecosystems.

The environmental provisions, although very general, have generated a potentially powerful mandate, and the Convention expresses strong concerns and common interests of states in environmental protection. Being a framework convention, States are free to deviate from it and may explicitly consider the interconnectedness between land and water, impacts from land use on water, the protection or preservation of specific freshwater ecosystems, and formulate thresholds pursuant to ecological factors.

It may be necessary to consider enforceable rights not only where environmental harm affects another watercourse state as envisaged in Article. A particular challenge vis a vis ever increasing human and economic uses is the release of a minimum in-stream flow of good quality water to maintain a river's biological, chemical and physical characteristics.

## The Rio Conventions and the Protection of Freshwater Ecosystems

The Convention on International Watercourses constitute a commitment and framework on behalf of countries to begin reversing the course of freshwater biodiversity's loss. When taken together with the three Rio Conventions, a transition to conservation and sustainable use of freshwater ecosystems can be followed.

### *The Convention on Biological Diversity (CBD)*

The CBD focuses on the integration of biodiversity concerns into development decision-making through the conservation and sustainable use of biological diversity and the fair and equitable sharing of benefits stemming from their utilisation. Albeit there are no explicit obligations in the convention relating to freshwater biodiversity, the CBD started action concerning inland waters at the occasion of the second Conference of Parties and adopted inland waters as a thematic area at COP-4.

The relevant work programme on biological diversity of inland water ecosystems promotes integrated water management; appropriate technologies; research, monitoring and assessment; and co-operation with other conventions and organisations. In management terms, work plans and programs adopted by the COP emphasise the need for integrated watershed, catchment and river basin management based on an ecosystem approach including transboundary water resources. Particularly the interrelation between water management and agriculture was stressed, as well as the need to use the ecosystem approach in all sectors which have an impact on inland water biodiversity. Further, the necessity to improve knowledge on the pressure to and status of biodiversity of inland waters is stressed.

It recommends to its parties to work out national reports on the status of freshwater biodiversity and plans for action, and to apply various measures, such as legal, administrative and incentive measures, strategic environmental impact assessment (SEA), awareness raising concerning alien species, genotypes and genetically modified organisms, transboundary co-operation and involvement of local and indigenous communities.

However, contribution of the CBD to further improvements in ecological-orientated water management is manifold. In particular, results of the ongoing attempts to improve knowledge on the ecological status of water resources and causes of the biodiversity's loss will be an important step towards the development of adequate counter-measures. Unfortunately, experiences in practical implementation is mixed and progress at the national level is slow.

For example, knowledge on status and trends of inland water biodiversity is still limited, due to the fact, amongst other things, that parties only slowly implement measures to assess, monitor and report. Case studies submitted by the parties are considered insufficient in number and detail. Further, concrete action is still small, despite the launching of a River Basin Initiative (RBI) aimed at the establishment of "a network to link and support activities and projects in which the principles and the practice of integrated management of biodiversity, wetlands, and river basins/catchments/watersheds will be demonstrated".

To progress the development of the initiative, the RBI has recently been awarded a UNDP project development fund grant. Initial work is under way and as a first step a "Country Needs Assessment" will be undertaken. But since the RBI was started only in 2000, results yet can not be evaluated. In general, since CBD's activities in the area of inland waters started rather late, the impact on practices in water management can not be assessed. However, there is a broad political consensus, and the measures that have so far been taken will lead to an enhanced diffusion of knowledge and good practices. Further, the CBD started a fruitful co-operation with the Ramsar Convention.

In accordance with recent CBD decisions, the CBD's Subsidiary Body on Technical and Technological Advice (SBSTTA) was instructed to work with the Ramsar Scientific and Technical Review Panel to adress issues of common importance (e.g. impact assessment, inland waters, alien species). Meanwhile "Joint Work Plans" were adopted and the Ramsar Bureau has proved to be a strong political actor.

The Ramsar Convention or "Convention on Wetlands" was the first modern global intergovernmental treaty on conservation and wise use of natural resources. Although its provisions are relatively straightforward compared to other international agreements, its focus in the beginning was rather narrow. But over the years, however, the Ramsar Convention has broadened its scope to cover all aspects of wetland conservation and wise use, recognising wetlands as ecosystems that are extremely important for biodiversity conservation in general and for the well-being of human communities.

One of its key obligation is to include wetland conservation in national land-use and water planning to promote "the wise use of wetlands in their territory". The principle of "wise use" is an important link to both the CBD and the Agenda. In recent years, the Ramsar Convention has adopted various documents and guidelines for taking due consideration to wetland conservation in water management. Recently, Ramsar is a pioneer in efforts to broaden co-operation and to harmonise work of the different Multilateral Environmental Agreements (MEAs). Beyond the co-operation with the CBD, the Ramsar Bureau has begun to develop direct co-operative efforts with the secretariats of other global treaties.

### *The UN Framework Convention on Climate Change*

Global climate change will impact on the world's freshwater

resources: With increasing global and regional temperatures, rainfall patterns will change throughout the world, and increases in evaporation rates, changes in the variability of water availability and altered demands for freshwater will occur. Increasing temperatures globally are likely to result in a warming of water temperatures in lakes and rivers.

While many uncertainties remain about the precise details of the regional impact on water resources, it is increasingly apparent that biodiversity will be threatened. Biodiversity in freshwater systems is distributed in a fundamentally different pattern from that in marine or terrestrial systems. Freshwater habitats are relatively discontinuous, and many freshwater species do not disperse easily across the land barriers that separate river basins. This has three important implications:

- Freshwater species must survive climatic and ecological changes in place.
- Freshwater biodiversity is usually highly localised, and even small lake or stream systems often harbour unique, locally evolved forms of life.
- Freshwater species diversity is high even in regions where the number of species at any given site is low, since species differ between one site and the next.

Because climate change is expected to occur very fast in comparison to the ability of freshwater ecosystems to adopt to it, a disequilibrium between species distribution will occur. This suggests that climate change would not only have a direct impact on the distribution of species, but that species could become extinct leading to a loss of biodiversity. In consequence, even if the ability of freshwater ecosystems to adapt to climate change highly depends on the rate and extent of expected changes, ecological goals of water management can no longer be achieved without taking climate change into account.

Climate change represents an additional threat to freshwater ecosystems and requires specific measures in water policy and concerted actions between international water policy and climate policy for common action. However, functional links between conservation of freshwater ecosystems and climate change are on two sides because climate change is also fostered by changes in freshwater ecosystems. Freshwater ecosystems provide services like the regulation of climate. Further, wetlands are considered significant carbon stores and carbon dioxide sinks. Hence, existing and additional wetlands might serve to stabilise the world's climate.

But since wetlands also are sources of greenhouse gases like methane and nitrous oxide, the net climate effects of the loss of wetlands or their restoration is still scientifically and politically controversial. Recent studies suggest that the net effect still is not fully exploited and obviously depends on the type of wetland, location, alternatives in land-use and other factors. However, since they are the world's largest carbon sink, in particular peatlands are considered very important and are the subject to controversial debates on carbon sinks under the Kyoto Protocol.

There are continuing efforts to enhance co-operation between the CBD, the UN Framework Convention on Climate Change (UNFCCC) and international water policy. The Framework Convention and its Kyoto Protocol are emerging as critical mechanisms to foster wise use of carbon resources and to implement a global trading mechanism for carbon credits. The inclusion of land use, land-use change, and forestry activities in accounting for net emissions under the Kyoto Protocol and in possible project activities carried out under the mechanisms of Article could promote conservation and sustainable use of certain freshwater ecosystems.

Peatlands still are excluded from the Kyoto Protocol, although its Article would provide for an agreement making land-use changes relating to wetlands part of the commitment pertaining to greenhouse gas emission. However, joint initiatives among the UNFCCC, the Ramsar Convention and the CBD provide important opportunities that are widely recognised but not yet fully realised.

## The Convention to Combat Desertification

While both the CBD and the UNFCCC clearly have linkages to floods, droughts, aquatic ecosystems, and carbon sinks, the UN Convention to Combat Desertification (UNCCD) has water and land management at its heart since its provisions are aimed at combating desertification and mitigating the effects of drought and desertification. Various assessments continued to point out that desertification results from complex interactions among physical, chemical, biological, socio-economic and political factors, that are of local, national, and global origin.

The protection and sustainable use of remaining freshwater ecosystems in regions which are affected by drought is certainly a key feature of strategies to combat or mitigate desertification. In many areas, implementation measures in water management focus their efforts on the protection of aquifers. Putting a light on the Convention, there are many objectives, principles and measures in common with the CBD and subsequent arrangements:

- *Objective*: Long-term integrated strategies that focus, inter alia, on improved rehabilitation, conservation and sustainable management of land and water resources.
- *Principles*: Integrated approach; establishment of a better understanding of the nature and value of land and scarce water resources
- *Measures*: Sustainable management of natural resources; sustainable agricultural practices.

## Interaction of Global Agreements

Considering functions and services of freshwater ecosystems as well as threats to their biodiversity, functional links with other key issues of international environmental policy are obvious. Rising awareness of these interdependencies have fostered increasing co-operation between the related conventions already adopted. The topic "biodiversity of freshwater" is tackled by a multitude of declarations, conventions and international organisations, those that have been mentioned and others both inside and outside the UN system (*inter alia* FAO, World Bank, Bonn Convention, Global Water Partnership, World Water Council, Wetlands International, IUCN).

In particular, with the CBD, the Ramsar Convention, the UNFCCC and the Convention on International Watercourses, there is an emerging global framework for action aiming to preserve the integrity of freshwater ecosystems. However, it is worth emphasising the different character and objectives of these conventions though: The character of the three Rio conventions is not comparable to that of the Watercourse Convention. The Rio conventions have a process character and are endorsed by an institutionalised follow-up (conferences of parties, secretary and boards, scientific support etc.) and a financial mechanism (Global Environmental Facility).

The Watercourse Convention, that was adopted after a long-lasting and controversial negotiation process and is still not ratified, is a legal framework for co-operation of riparian states. This is probably due to the fact that agreements are subject to regional negotiations. Finally, the Agenda, water chapter, is a political declaration what means inherent limits in effectiveness. The CBD's and Ramsar Convention's subject is the preservation and sustainable use of freshwater biodiversity and wetlands respectively.

In the negotiation process of the UNFCCC preservation of wetlands only plays a minor part, albeit political controversial. Further, the subject of discussion is not biodiversity of wetlands but their net climate effect. There might be concrete cases where maximising climate effect and preservation of biodiversity are conflicting goals though. The Desertification Convention also does not directly tackle freshwater biodiversity, even if the preservation of freshwater ecosystems is in many affected regions a key prerequisite to reach its objectives.

In turn, the Watercourse Convention focuses more on water as a resource subject to beneficial human and economic uses, and on the establishment of rules and procedures to balance competing interests of riparian states. However, both the principle of "sustainable utilisation" and the environmental provisions provide a potentially powerful mandate for taking due consideration of watercourses' ecosystem protection. Further, State sovereignty over natural resources is still the ground upon which the Watercourse Convention is build while the preamble to the CBD introduces a somewhat different concept that limits States sovereignty when it states that the "conservation of biological diversity is a concern of humankind".

The CBD concept does not affect the "ownership" dimension of States over their resources but the "use" dimension. The implications, however, are far reaching if rights were assigned to non-riparian States to the degree that is necessary to address the "common concern". To prevent global interference in national and regional water management, the concept may be modified into "regional common concern". Rights generate obligations that might be differentiated according to economic strength of co-riparian and non-riparian States, and financial transfer might be considered if States lack adequate resources.

However, at the global level there is a fruitful co-operation of the CBD and the Ramsar Convention to foster the topic "freshwater biodivesity". The effectiveness of their work and the advances reached (e.g. work programme, River Basin Initiative, indicators) can be traced back to the fact that negotiations take place in a de-politicised forum of experts. Even if progress in implementation is rather slow and translating the holistic approach of the CBD into concrete action in the light of incomplete scientific knowledge has proved difficult, there is now a stronger focus on implementation.

Further, CBD and Ramsar have initiated a co-operation with many other actors relevant in international water policy (World Commission on Dams, World Bank, IUCN) and the conventions mentioned. Albeit not perfect, this emerging network has intensified scientific and political discussion and can potentially lead to more coherent actions on the global level. Further, this network favours a convincing policy style of mutual learning, diffusion of best practice, monitoring and reporting.

There are many complaints concerning confusion and frictional losses due to the multitude of actors and conventions in international environmental policy, but the co-operation in the field of biodiversity orientated water policy proves that these institutional obstacles can partly be overcome. For example, the Joint CBD-Ramsar Work Plans have become model by which other conventions are seeking to develop complementary working arrangements with the CBD. In general, the weakness of international policy relating to freshwater biodiversity is less the result of lacking global co-operation but can rather be traced back to less political support at the regional and national level.

## THE PRINCIPLES OF ECOSYSTEM MANAGEMENT

Even if some synergies between the different international arrangements and processes relating to the biodiversity of freshwater resources still deserve more attention and none of the international treaties and institutions already deals fully with freshwater ecosystems, there is an increasing consensus on the highest political levels to manage water, water-related processes, and ecosystems in a sustainable manner.

However, at the level of implementation the value of ecosystems is rarely taken fully into account. In order to close this implementation gap the contracting parties of the CBD and other advocates have proposed a so-called "ecosystem-based approach" as an appropriate way of managing natural resources.

## THE ECOSYSTEM APPROACH

The ecosystem approach is a strategy for the integrated management of land, water and living resources that promotes conservation and sustainable use in an equitable way. The term ecosystem is used more as a mental construct suggesting complexity and systems interaction rather than a geographic entity. An emphasis on the complexity of system-wide interactions highlights scientific uncertainty and results in the need to deal with this uncertainty explicitly by acting conservatively and managing adaptively: setting a course of action based on a set of hypotheses, monitoring what happens, and re-evaluating the direction on what one learns.

At its most fundamental level, an ecosystem approach maintains diversity as a means of building resilience against catastrophic events in biological, economic, organisational, and political systems. At its second meeting in 1995, the Conference of the Parties of the CBD adopted the ecosystem approach as the primary framework for action, and subsequently has referred to the ecosystem approach in the elaboration and implementation of the various thematic and cross-cutting issues work programs under the Convention. The issues concerned include:

- Biological diversity of inland water ecosystems.
- Marine and coastal biological diversity.
- Agricultural biological diversity.
- Forest biological diversity.
- Indicators of biological diversity.
- Incentive measures and environmental impact assessment.

By recommending the ecosystem approach as the guiding concept, in May 2000 the fifth COP took a next step towards its practical verification in the frame of the CBD.

Concerning inland water ecosystems a work programme was adopted by the COP which highlights the importance of designing integrated watershed, catchment and river basin management strategies. Particularly the interrelation between water management and agriculture was stressed, as well as the need to use the ecosystem approach in all sectors which have an impact on inland water biodiversity.

The ecosystem approach was initially developed as a research paradigm and not intended to serve as the basis for resource management. With the political application of the notion "ecosystem-based approach" advocates intended to find an adequate language in response to the shortcomings of classical nature conservation approaches.

Although classical nature conservation (e.g. designation of protected areas) can be considered successful in many regions and also is an important module in ecosystem management, its recognition of human needs and other legitimate sectoral interests was rather insufficient. It did not put enough emphasis on the biological diversity outside protected areas. Furthermore, the primary method of protecting freshwater biodiversity has been to designate particular species as threatened or endangered making them subject to national recovery programs or international protection.

Unfortunately, this approach is failing and it demonstrates the need for a more holistic approach in order to overcome the shortcomings of traditional efforts in nature protection policy. In the United States, for example, no aquatic species has ever graduated from the government's endangered species list, but 10 species of fish have been removed due to extinction. Equally, ecosystem management responds to the obvious shortcomings of classical single-resource approaches focused on exploitation.

## THE 12 MALAWI PRINCIPLES

- Management objectives are a matter of societal choice.
- Management should be decentralised to the lowest appropriate level.
- Ecosystem managers should consider the effects (actual or potential) of their activities on adjacent and other ecosystems.
- Recognising potential gains from management there is a need to understand the ecosystem in an economic context. Any ecosystem management programme should:
  - Reduce those market distortions that adversely affect biological diversity.
  - Align incentives to promote sustainable use.
  - Internalise costs and benefits in the given ecosystem to the extent feasible.
- A key feature of the ecosystem approach includes conservation of ecosystem structure and functioning.
- Ecosystems must be managed within the limits to their functioning.
- The ecosystem approach should be undertaken at the appropriate scale.
- Recognising the varying temporal scales and lag effects which characterise ecosystem processes, objectives for ecosystem management should be set for the long term.
- Management must recognise that change is inevitable.
- The ecosystem approach should seek the appropriate balance between conservation and use of biological diversity.
- The ecosystem approach should consider all forms of relevant information, including scientific and indigenous and local knowledge, innovations and practices.
- The ecosystem approach should involve all relevant sectors of society and scientific disciplines. These principles were agreed at CBD's COP-5 under decision V/6 with minor modifications.

Therefore, ecosystem management is proposed as a modern way of managing natural systems and generally intends to use natural resources within their natural limits. However, although there are still some controversies on how to translate ecosystem management into concrete management terms, parties of the CBD have already agreed on basic principles to guide the *modus operandi*.

A key point is that the ecosystem approach is a holistic process for integrating and delivering in a balanced way the three objectives of the CBD: conservation and sustainable use of biological diversity combined with equitable sharing of biodiversity's benefits. In particular, the agreed principles of the ecosystem approach stress the importance of:

- Societal choice concerning management objectives.
- The "economics" of biodiversity (internalisation of costs and benefits, balance between conservation and use).
- Adequate institutional structures (decentralisation, management at the appropriate scale).
- Adaptive management which enables policy makers to anticipate and cater for the natural dynamics of ecosystems.
- Consideration not only of scientific knowledge but also of indigenous and local knowledge.

In order to translate this rather abstract principles into guidelines for water management it is first necessary to define the appropriate management level to meet the general objectives. While the cited principles stress the importance of decentralised approaches in general terms, the idea of river basin management is considered an adequate starting point for the implementation of the ecosystem approach in water management.

The reasoning is that river basins offer many advantages for strategic planning and that they represent the most appropriate geomorphologic units on which to base ecosystem management practices. However, concerning ecosystem protection there are some limits: groundwater catchments only rarely match with surface water catchments; water quality and quantity inside the river basin interacts with other influences outside the catchments boundaries; estuarine ecosystems, for example, can only be sufficiently protected by taking into account the interaction of freshwater and tidal currents.

Therefore, river basin management should not be perceived as a panacea but it does provide a sound geographical basis for implementation. Briefly, ecosystem management should be an integral component of comprehensive river basin management strategies. Hence, the explicit switch to ecosystem management leads to an accentuation of the main focus in water management because ecosystem management demands:

- A broader perspective: whole ecosystems and the full range of services, no sectoral view.
- Efforts to balance natural ecosystem robustness and human-induced alterations.
- Collation of information on the status of, and threats to, the biodiversity of the freshwater ecosystem concerned (including impacts of the introduction of alien species).

- Explicit consideration of land-use, in-stream flow and water temperature for maintaining ecological functions.
- A long-term perspective.
- Consideration of the full economic value of freshwater ecosystems (include its biodiversity).
- Recognition of ecosystem's change as a matter of fact (i.e. restoration of damaged or destroyed ecosystems will never lead to the "pristine" state; adaptive management to deal with uncertainties).
- Integration of scientific knowledge and modern management tools (Decision Support Systems, Geographical Information Systems etc.) into decision-making.
- Identification of win-win options between protection and use of freshwater ecosystems: "use them, but don't lose them".
- Consideration of the full range of policy instruments applicable in water management (economic and legal instrument, co-operation etc.)

As most of these elements are widely agreed upon the rhetoric of international water policy, the next steps must include attempts to tackle the implications of the ecosystem approach in practice. For example, only little is known about the value of ecosystem management in political terms for at least two reasons:

- The concept appears highly demanding because much information – sometimes on "invisible" relationships between human-induced pressures and natural reactions – is needed, and effective vertical and horizontal co-ordination of sector policies and planning instruments has to take place. Today, even in wealthy nations a lack of integration across sectors in managing water resources can be identified.
- It is necessary to decide which services of water receive priority, and how the costs and benefits are to be distributed. The developed world, for example, can "afford" to stop or to remove development projects in favour of ecosystem integrity (e.g. Great Whale Hydropower Project in Canada, Edwards Dam in the US) and to use sometimes generous subsidies to compensate land owners for restricting their use of the lands. The restoration of river ecosystems in the developed world also is an example for costs emerging in the short terms. Although long-term economic benefits are regularly provided, political decision are often dominated by short-term considerations.

## ECOSYSTEM MANAGEMENT IN PRACTICE

In order to discuss shortcomings of current approaches to water management from an ecosystem perspective and to identify cases of "good practice" which could offer a guide, we stress three key dimensions of ecosystem-based water management, i.e. the international, the institutional, and the instrumental dimension.

## INTERNATIONAL CO-OPERATION AT TRANSBOUNDARY FRESHWATER ECOSYSTEMS

Since many basins cross political borders, international co-operation is of prime importance for the implementation of an ecosystem approach in river basin management. In general, water sharing between riparian states of the Ganges, Nile, Jordan or Mekong, for example, is an important political and strategic issue for the states concerned. Further, it is claimed that diverging interests of states would not only be obstacles for the integrated management of transboundary rivers, but would also be source of conflicts in the next years.

Although this pessimistic view appears foreshortened, diverging political and strategic interests of riparian states might prevent common action to ecosystem management at transboundary rivers and lakes. However, there are many examples of co-operation between riparian states in the allocation and control of freshwater resources. The United Nations has compiled a list of 3,707 agreements. Issues of quantity and quality of water resources have been resolved in some, but not all, cases.

In general, cooperation is confined to a limited scope (e.g. electricity, fishery, quantity and quality of water) while the governments involved only rarely commit themselves to an holistic ecosystem approach. Further, Wallensteen and Swain suggest that many of the bilateral agreements that have been developed in the past are under threat because of the increasing demands placed upon water resources through growing populations and the need for food security.

Tensions over water exist in river basins and there are few water-sharing agreements that includes all riparians or ensures full protection of the freshwater ecosystems. Thus, obstacles for the implementation of ecosystem management are obvious and stem from a lack of common understanding, adequate institutions (international, national and sub-national), and mechanisms for conflict solution and sharing of costs and benefits. Yet, there are some encouraging examples that demonstrate potential gains.

The experience of Canada and the United States in governing the Great Lakes ecosystem is often cited as a model. Although the success story "Great Lakes" is clearly linked to its particular political, economic and social context and its international transferability is limited, a number of characteristics might have relevance in other contexts. Valiante et al. stress as key prerequisites for successful co-operation at the Great Lakes inter alia:

- Binationalism in formal mechanism for fact finding on contentious issues.
- Formal structures: The International Joint Commission and newer arrangements have provided a depoliticised forum for resolving potentially divisive environmental issues.
- Equality and parity in structure and obligations.
- Flexibility to adapt to changing circumstances.
- Substantive objectives of the regime that provide progressive direction and flexibility; and most importantly.
- Development of an influential and informal binational "community".

## THE GREAT LAKES EXPERIENCE IN ECOSYSTEM MANAGEMENT

The Great lakes basin is a significant international ecosystem. Shared between Canada and the United States, it contains approximately 20 percent of the Earth's fresh surface water and is home to some 35 million people. Environmental pressure is also a characteristic of the Great Lakes, i.e. eutrophication, contamination from toxic substances, fluctuating water levels, introduction of alien species, loss of native fish species and wetlands.

Despite still existing pollution and deficits in implementation in the two countries (different paces, partly non-compliance), Canada and the US have created an effective institutional arrangement for implementation of an ecosystem approach: The International Joint Commission - a longstanding, binational, generalist organisation assisted by several newer binational bodies with different tasks.

Substantial progress was made towards ecological goals. Further, co-operation of Governments has changed behaviour towards the entire ecosystem. Significant resources were devoted to studying the ecosystem and to carrying out clean-up programs and restoration measures. Collaboration with industries was effective and broad participation (states and provinces, environmental and citizen organisations) was ensured. Despite serious disagreements at times, the two countries as well as the sub-national political entities have continued to participate in the Joint Commission.

However, it is worth mentioning that international co-operation in comprehensive river basin management usually takes a very long time to develop and often starts with less demanding arrangements than a more holistic ecosystem approach. The establishment of the International Commission for the Protection of the Rhine (ICPR) took place in 1950, but common measures to reduce the increasing pollution of the river were not agreed upon until the 1970s.

Further, much progress was achieved in the aftermath of the Sandoz disaster in 1986. In reaction to the disaster riparian states agreed on the Rhine Action Plan, a comprehensive plan to tackle environmental problems in a holistic manner including water quality aspects and the protection and restoration of habitats and floodplains. Further, in 1998 riparian states adopted an Action Plan for Flood Defence which includes, apart from water management, measures in the fields of nature protection policy, agriculture and forestry and spatial planning.

Today, the ICPR serves as an example for good practices in ecosystem management of international river basin because it explicitly considers biodiversity aspects and the interaction of land-use development and water management. However, one key lesson from the ICPR is that non-legally binding action plans are worth considering in international co-operation because they can be prepared more quickly and can be more ambitious as legally binding treaties.

The co-operation on the River Rhine has also shown that the identification of a flagship species can serve as a symbol for ecosystem management and might help to mobilise public support, i.e. efforts to develop a comprehensive ecological management approach were heavily stimulated by the programme *Salmon 2000* started in 1991. However, it was the combination of the Action Plans, the emergence of mutual trust, owing to the co-operative atmosphere in formal organisational structures, and the provision of considerable financial resources which made the ICPR a success story.

There are also some encouraging examples for improved consideration of environmental and biodiversity aspects in the developing world, even if the label "ecosystem approach" is rarely applied. The Mahakali Treaty, for example, explicitly addresses issues of river ecosystem maintenance in Uttar Pradesh. Even if economic aspects of water use and sharing of benefits are dominant, co-operation between India and Nepal will also lead to an amelioration of the river basin in environmental terms.

Further, the conflict between India and Bangladesh exemplifies the possibility for common action to tackle the problem of over-extraction leading to reduced availability of water resource in the downstream riparian state. In that case, a barrage was built 18 kilometres from the Bangladesh border on the Ganges River to divert water through a feeder canal to supplement the dry season flow of the Bhagirathi-Hooghly, the river that serves Calcutta and irrigation needs in West Bengal. On occasion this had reduced the flow of the Ganges substantially, causing damage to the riparian environment, and in particular affecting the world's largest mangrove forest, the Sunderabans.

A number of attempts had been made to resolve this dispute, culminating in the signing of the treaty between the two countries in 1997 that defined the mechanisms for sharing the Ganges waters. Other examples demonstrate that the development of a co-ordinated water resource development and management scheme is crucial. This is often the major aim of harnessing riparian states, as borne out by many cases in the past. A realistic basin-wide management scheme may be developed only through the collaboration of riparian states.

Having such a scheme serves as proof for other countries, in particular donor countries and aid agencies, that riparian countries are on good terms, to the extent that they have developed the scheme through collaboration and are ready to implement it in a co-ordinated manner. An example is the development scheme of the Mekong River Basin in South- East Asia, although many problems could not be solved. The management of the Okavango Delta and its biodiversity rich wetlands also has improved albeit increased water withdrawals for urban use is still a politically controversial issue and has threatened significantly biodiversity in the past.

Further, the Lake Chad Basin countries are actually undertaking a project for the sustainable management of groundwater, wetland and land resources in the basin. In addition, transboundary conservation areas (TBCAs) provide a mechanism to promote sound ecological management of transborder ecosystems, and, at the same, promote opportunities for regional political, economic, and cultural co-operation. Ecological benefits of TBCAs are mainly associated with protecting larger areas. Specifically, TBCAs may improve the protection of internationally shared resources, such as watersheds, that provide important ecological and economic services.

TBCAs may encompass critical portions of watersheds for multiple purposes, such as pollution control, clean water supply, wetland management, and carbon sinks. Hence, TBCAs might serve as a core zone for integrated management in entire river basin. Recently, such transfrontier conservation projects are taking place on East African Lakes (e.g. Lake Victoria, Lake Malawi), for example, which are globally important for biodiversity as centres of evolution for African freshwater fish. Similar actions are occurring between South Africa and its neighbouring countries. In all these cases, joint international recognition of transboundary conservation projects, such as World Heritage Sites and Peace Parks, might strengthen co-operation. Because many of the approaches and projects mentioned are still in their early stages, it is impossible to judge their success in terms of ecosystem management. But even if progress can be expected, rapid changes will rarely occur because of the lack of adequate institutional capacities in many countries concerned.

This means that international co-operation might favour more comprehensive and ecosystem-based approaches in water management, but as vulnerability and degradation of freshwater ecosystems are mostly rooted in incomplete or inappropriate legal and policy frameworks at the national level the pace in implementing ecosystem-based strategies heavily depends on domestic policy reforms. Different stages of economic development and diverging "philosophies" in water management (e.g. water rights, pricing) make ecosystem management a long-term and complicated process.

As success stories have proven effective ecosystem management of transboundary freshwater ecosystems is usually part of a broader evolutionary process of building capacity towards co-operative water resources frameworks.

## NATIONAL INSTITUTIONS AND ECOSYSTEM MANAGEMENT

A key prerequisite for successful implementation of an ecosystem-based approach in river basin management are adequate institutions that reach the full range of uses and users that are connected in freshwater ecosystems. While the need for integrated and cross-disciplinary approaches to land and water management are sufficiently described in theory, too little is known how the demanding concept could succeed in practice.

### Local/regional Government Boundaries

Freshwater systems extend over a wide range of administrative and bureaucratic boundaries. It is rare that a single management unit has the full responsibility for the management of a freshwater ecosystem. Therefore, the failure to achieve integrated and environmental-orientated management has been attributed to the strength of local governments and sectoral orientated administrations. This is obviously confirmed by many cases of sector-by-sector fragmented approaches which remain the norm.

Today, there are at least some encouraging examples that prove that ecosystem management of water resources can be fostered by political measures. The new approach of the European Union (Water Framework Directive) is a frequently cited example for an institutional innovation that will help to strengthen the integration of ecological objectives in water management, basin by basin. Obligations for the member states of the European Union not only encompass the establishment of integrated river basin planning as a key instrument but also the consideration of the effects of water use on aquatic and related terrestrial ecosystems.

Although there is still a main focus on water quality, habitats and protection areas, for example, are an important part of the management plans to develop. Equally, the Tennessee River Basin, the new Brazilian water law and the new South African water law show promise for implementing an ecosystem-orientated management.

## ECOSYSTEM-ORIENTATED MANAGEMENT IN SOUTH AFRICA

The Mgeni Catchment in KwaZulu-Natal, South Africa, is the most socially and economically important catchment in the region. However, water demand in the catchment is fast approaching the limits of water availability, and water quality is deteriorating. The need to manage the water resources in the Mgeni Catchment holistically has led to the formation of Mgeni Catchment Management Plan, the objective of which is to ensure that water resources in the catchment are managed in a sustainable way.

These efforts on regional scale are heavily supported by the recently adopted new National Water Act, which focuses on the integrated management of water resources on a catchment basis and explicitly addresses the integration of ecological important areas. The ecological and economic importance of wetlands in the Mgeni catchment has been recognised, and in support of the new water law and the Catchment Plan, plans has been initiated to integrate wetlands and water resources management. To this end, a collaborative effort involving water management institutions, conservation organisations, wetland experts and landowners aimed at the rehabilitation and conservation of priority wetlands in the Mgeni Catchment has been started.

Administrative co-ordination is a challenge in all countries but they are particularly grave in federal political systems. Against this background, there are two models to distinguish in order to classify practical approaches. In the "authority model" water management is a seperate policy sector, organised on the basis of hydrological boundaries. Since the advantages of this approach (e.g. facilitated basin-wide co-ordination of activities) are obvious, shortcomings may occur due to insufficient co-ordination with other sector policies and financial instruments.

In the "commission model" administrative bodies deal with a range of water-related policies (environmental management, land-use planning etc.) but they do not fit with hydrological boundaries. Therefore, basin-wide co-ordination is handled by means of intergovernmental co-ordination and co-operation, often in the frame work of river basin commissions and endorsed or enforced by central government interventions. In practice, these two models have different advantages but both can serve as a tool to enhance ecological considerations in water management.

## THE AUSTRALIA'S MURRAY-DARLING RIVER BASIN

The case of Murray-Darling basin demonstrates how obstacles stemming from multilevel government might be tackled. Ecological degradation of the river (pollution, salinity, water shortages, degradation of freshwater ecosystems) created a strong impetus for action. Albeit there are still some barriers to interstate river basin management, and a fully integrated approach for wetland protection in the whole basin is still lacking, marked improvements in implementing an ecosystem-based management could be achieved in recent years.

Reasons for success for improved inter-state river basin management are manifold and can be put down to effective interaction of measures at the level of the Australian Government and coordinated efforts of the states concerned. While the existence of an authority with basin-wide responsibilities including an executive arm (Murray-Darling Basin Commission) serves as an example for an institutional arrangements with basin-wide responsibilities, environmental orientated reforms, adopted by the council of Australian Government, have pushed the inter-state co-ordination for enhanced integration of ecological aspects.

Henceforward, a special allocation of 25% of flow is being set aside for in-stream ecological purposes and at the same time water charges are increased, water and salinity reduction markets are created and nutrient reduction measures are being implemented. The agreement of the Murray-Darling Basin Ministerial Council in 1995 to establish the collective cap on water extraction from the river represents a unique inter-state initiative to reduce water stress and control water salinity.

This action was enforced via an already existing system of volumetric water allocation across regions, sectors and individuals.

The Malawi Principles rightly emphasise the necessity for decentralisation because in the past water management was over-centralised in many countries. But devolution of responsibility for water management and allocating planning to local or regional entities is not a self-evident process because of the many political obstacles and diverging interests. Further, if decentralisation of water management has occurred, there is regularly a need for the development of appropriate mechanisms for the resolution of conflicts.

While in some cases more or less effective institutions to overcome these obstacles have been created, the majority of countries still lack appropriate arrangements. The still insufficient allocation of rights to water supplies from the Cauvery River between the states of Karnataka and Tamil Nadu in India might serve as a recent example with serious political and environmental consequences. These problems are made worse because the states concerned have no established platform for collaboration and co-ordination.

Also in the developed world major problems in the co-ordination of the interests of different government levels still exist, as the case of conflict between the national government of Canada and the province of British Columbia concerning the realisation of benefits from storing water and producing hydroelectric power exemplifies. However, from the viewpoint of ecosystem management, adequate institutions to deal with different regional or local interests must be developed. As successful practical approaches demonstrate, an active and enabling role of the central government is an important factor.

In other words, decentralisation alone will not automatically lead to an effective ecosystem management because crucial factors for success remain external from the viewpoint of regional or local ecosystem managers, e.g. sector policy measures, provisions for assignment and enforcement of property rights and further legal instruments, rules for the identification and designation of protected areas, rules for conflict resolution, inter-regional transfer schemes in cases where benefits of ecosystem protection mostly occur elsewhere, but not or only partly in the region or river basin concerned.

## SECTORAL INTERESTS

Today obstacles for an effective implementation of the ecosystem approach in water management often result from the characteristics of many national (or regional) water management institutions that are typically concerned with one set of issues and that do not effectively represent the interests of all constituencies in their operation. In general, water management institutions still tend to be technically oriented and are bureaucratic bodies based on one or, at best, two specific aspects of water management.

They seek to maximise their supply or outcome with, in many cases, little regard for the social impacts and the integrity of the whole ecosystem. As reported by Kottelat/Whitten the conservation of endangered fish species, for example, often suffers from being placed under different governmental bodies (conservation and fisheries agencies). Since there is no effective co-ordination between the agencies, legally protected fish species are threatened and exotic species are introduced and destined for forms of aquaculture even though they harm local biodiversity.

In many cases, even environmental or conservation agencies are not really engaged in ecosystem management, but concentrate their efforts on particular species or habitats that are easily to observe and receive the most publicity. In general, management units act independently, and have incentives to maximise the welfare of their own members and are often unaware of the effects of their activities on other users of the ecosystem, especially with regard to impacts on certain functions of freshwater ecosystems, such as the support of the biodiversity in an entire river basin.

But if management units are unable to incorporate these effects, the whole water management systems will not lead to an efficient allocation of water in terms of ecosystem management. Experience suggests that there are some effective approaches to attenuate these obstacles:

- Organisational reforms in administration, e.g. the consolidation of water administration and environmental units or basin-specific inter-ministerial committees.
- Development of sector biodiversity programs and plans, e.g. biodiversity action plans in fishery, water policy etc.
- Strategic Environmental Assessment (SEA) including effects on biodiversity as a means of enhancing consideration of ecosystem management in sector-orientated management units.
- Economic incentives.

Even in many river basin institutions that are entitled to tackle cross-sector management aspects, certain values of freshwater ecosystems have been traditionally over-represented (e.g. agriculture, industry) while other users and user groups as well as ecological concerns were under-represented. Progress will be made if all user groups and the public are involved in management processes at the river basin level. This means that participation and involvement of stakeholders are crucial for the implementation of ecosystem management.

An interesting approach are "River Parliaments", i. e. a forum in which all interests are represented, and which defines the rights and responsibilities of different parties over the management of the river. Although practical experiences are limited and it will be extremely difficult to get such an initiative going let alone organising the representation of non-use values of freshwater ecosystems, the approach is based upon social rather than technical processes and as such might offer opportunities for ecosystem management.

In order to facilitate public participation it is useful to acknowledge the dual character of institution building, as shown in the slogan "Think basinwide, act watershed specific". Certain decisions can only be effectively made at the basin level, while in many cases other decision and effective integration of stakeholders in decision-making could be organise more effectively at a much lower level such as the sub-catchment.

## DECENTRALISATION, LOCAL COMMUNITIES AND LIVELIHOOD CIRCUMSTANCES

Since river basins can be seen as an appropriate level for action, ecosystem management needs both decentralisation and a kind of hierarchy on the basin level to provide overall direction in the water management process. Strategic planning defines broad scale regional goals and basin-wide resource utilisation and management plans to mobilise the next level in the hierarchy. At this lower level, catchment management plans are translated for specific sub-catchments.

The lowest level of the hierarchy is based on operational decisions for, inter alia, specific river reaches, reservoirs, wetlands or to specific lands. But even if certain hierarchical structures are useful to effectively implement ecosystem management, especially the potential role of local communities is often underestimated. Since local communities, in particular in developing countries, are directly affected by the degradation of freshwater ecosystem, mobilisation of their knowledge and the consideration of traditional rights and uses is a key element to broadening the water management perspective and to take potential economic losses of ecosystem's degradation into account.

In many cases, the search for secure local livelihoods forces local communities to degrade natural resources in the course of economic activities. Therefore, ecosystem management will frequently not be successful if local livelihood circumstances and traditional rights and uses are not taken into account. It is equally important not to idealise local approaches because there are often rather single-issue institutions, and water user associations, for example, may be effective when the efficient allocation of goods and services provided by freshwater ecosystems requires intrasectoral rather than intersectoral reallocation of water.

But there are many cases which demonstrate the potential benefits of local participation in ecosystem management strategies and underline the need to achieve both regional co-ordination and local flexibility. For example, in Pohnpei, a larger volcanic island that belongs to Micronesia, a community-based approach in watershed management started in 1994 which is an effective learning process and it appears to be the only alternative in view of diminishing external aid and restricted government capacities.

Local institutions might be particularly important in cases of extreme environmental events. Howorth emphasised the importance of local social organisations in preserving collective goods and maintaining the integrity of freshwater ecosystems in Burkina Faso. Pirot et al. describe the importance of involving local communities and using their knowledge for the success of the Hadejia-Nguru Wetlands Project in Northern Nigeria.

## CHANGING ROLE OF THE STATE

New trends in water management force a reconsideration and readjustment of the role of the state in water management. They demand not only that states, in particular central governments, withdraw from many operative activities but, that it take on many new regulatory tasks, often of a very different nature and requiring different skills and knowledge of the public sector officials. The ecosystem-based approach is a good example for these new tasks.

There is a strong tendency to enhance the role of private enterprises in water management strongly reducing the role of the state in water management and traditional tools of governmental actions are not longer applicable. Since there are many services and goods provided by freshwater ecosystems which directly benefit users, privatisation might be useful to enhance economic efficiency. But it is worth noting that certain services of freshwater biodiversity have a clear public good character.

Therefore, effective ecosystem management demands that privatisation of public utilities, for example, is accompanied by effective regulation for those environmental services (e.g. protection of species and their habitats, flood control) for which privatisation and the establishment of private property rights is not possible. This is clearly indicated by international experiences of privatising water utilities. However, today governmental agencies alone are in many cases unable to plan and enforce ecosystem management.

## THE ROLE OF INCENTIVES IN ECOSYSTEM MANAGEMENT

A successful ecosystem management strategy must integrate environmental, social and economic factors. However, there are few practical methods for directly linking these different aspects of water management, nor is there one "perfect" instrument that fits all uses and related environmental consequences.

### How to Overcome Lack of Knowledge and Awareness?

What is needed as a first step is an understanding of cause-and-effect linkages between water use, land-use patterns, hydrological regime and biotic response. While some threats to freshwater ecosystems are quite visible and anticipated, e.g. impacts of dams, drainage canals etc., others are not, and corrective action is only sought when degradation has already entailed significant threats. For example, pollution may be sometimes overlooked due to the seemingly unrelated nature of pollution sources and affected ecosystems.

Furthermore, effects of the introduction of exotic fish species are sometimes hardly to assess in advance because of the generally poor quality of available taxonomic data and a lack of knowledge concerning long-term results. Equally, the spatial interconnectedness of water sources and land use in the catchment areas are often not well understood. Threats to wetlands such as floodplains, for example, result from changes in chemical, physical, and hydrological balances occurring from activities that are often located at a distance along the upper catchments of the river systems.

Further, unfamiliarity with the full range of freshwater biodiversity also leads the public and water planners to focus on those aspects of biodiversity that directly interact with local and commercial activities while the large-scale dynamics and complex interactions are neglected. Therefore, ecosystem management should start with the identification of the main characteristics of and pressures on freshwater ecosystems, especially for the most fragile ones.

Even if scientific knowledge is still limited, there are many tools that may help to better predict consequences of human actions on ecosystems, e.g. simulation models, environmental impact assessment, geographical information systems, decision support systems etc. These tools might provide useful information for decision-makers but in many cases their application is limited due to more fundamental problems in administrations: lack of staff, lack of training and awareness, insufficient financial provisions, and lack of community involvement.

In addition, better knowledge of the status of and threats to freshwater ecosystems is crucial in political terms because it provides a necessary basis for a common understanding and the identification of common objectives of all participants and groups or individuals affected. Adequate provision information and free access to it also means empowerment of people (e.g. third party rights to challenge decisions, rights of access to information) whose engagement is fundamentally needed in order to adequately ensure consensus and push ecosystem management forward. This was proved, for example, in the case of proposed structural engineering in the Pantanal (Brazil) which was drastically reduced in reaction to the outcomes of an environmental impact study and participation of civil society and NGOs in the planning process.

The identification of pressures and threats should integrate the assessment of economic effects of the loss of biodiversity. Such economic estimations are necessary not only for better prediction of social and economic impacts but also to answer the crucial question of how much nature and biological diversity in a river basin should be preserved or might be "sacrificed" from society's point of view. There is a growing number of studies which demonstrate how the full economic value can be assessed and used in water management decision-making.

*Economic Benefits of Freshwater Ecosystems*

---

*Direct values*: Fishing, fuel wood, building poles, thatch, hunting, wild foods, medicines, agriculture, pasture, transport, recreation etc.

+

*Indirect values*: Water quality, water flow, water purification, groundwater recharge, flood control, storm protection, nutrient retention, micro-climate etc.

+

*Option values*: future pharmaceutical, agricultural and industrial applications of biological resources, leisure use, water based development.

+

*Non-Use values*: Intrinsic, existence, cultural, aesthetic etc.

=

Total Economic Value

---

For example, for the US, total direct economic losses (e.g. commercial fisheries) and losses of public goods by wetland users (recreation, recreational fishing, waterfowl hunting) associated with wetland losses from 1954-1992 were estimated at US$ 136 billion, or an imputed average of US$ 10,558 per acre of wetlands lost. Adding the estimate of non-user values to this brings the total to US$ 598.2 billion, or US$ 46,556 per acre lost since 1954. There are similar results in valuation case studies for developing countries.

Further, although there are practical difficulties, economic valuation is also useful in assessing benefits provided by international important freshwater ecosystems that have an economic value for persons thousands miles away. This fact is explicitly addressed in the CBD in its Articles 20/21. It is not, however, necessary to conduct "perfect" economic valuation to realise that economic aspects of freshwater ecosystems are important and that it is essential to recognise the costs and benefits in decision-making.

As Emerson has reported, the most important value of urban wetlands in Nakivubo (Uganda) is the role that they play in assuring urban water quality, but equally poor urban population heavily depends on wetland resources for food and income. This case demonstrates that economic valuation is not just a place of scientific interest but also that the recognition of benefits of the method is especially important in the frame of economic development and in strategies to provide income for the poorest communities.

Since in many cases degradation occur due to the conversion of natural ecosystems to way of economic utilisation (agriculture, settlements), ecosystem managers regularly need economic arguments and concepts to overcome social and economic obstacles to determine which protective measures are necessary.

## HOW TO ENCOURAGE PEOPLE?

Generally, it is widely agreed that the multitude of threats to inland water systems demand a variety of instruments to preserve an ecosystem's integrity. As experiences in many countries have demonstrated, regulatory instruments ("command-and-control policy") might be adequate to reduce pressures on freshwater ecosystems in some cases. For example, there are instances of countries which applied effective regulatory instruments (e.g. emission limits) in their overall strategy to enhance water quality and herewith produced positive side-effects on the preservation of freshwater biodiversity.

In addition, mitigation and compensation measures might help to reduce the biodiversity impact of plans and projects. But regulatory instruments show various shortcomings: non-compliance occurs regularly and improvement of the ecological situation is achieved at a high price. Furthermore, they do not sufficiently take into account the probably most important underlying force of threats to freshwater ecosystems: the lack of adequate incentives to take the full economic and social value of freshwater ecosystems into account.

Therefore, both the CBD and the Ramsar Convention, have called for incentive measures to preserve freshwater biodiversity. Equally, Agenda has recognised the need to develop economic incentives to protect the environment. In the past policy failures and market failures in conjunction with the already mentioned institutional failures have generated a lack of adequate incentives: Policy failures occur if the causes of degradation of freshwater ecosystems are not recognised or not sufficiently tackled.

Efforts to conserve wetlands via regulatory instruments or management programs might fail because more fundamental problems of policy intervention failure (e.g. agriculture subsidies, low water prices) are not taken into account. The underlying reason for market failures is that economic activities are undertaken without consideration of the full impact of the activity on the integrity of freshwater ecosystems. Here, market prices do not fully reflect the economic value of water resources. Therefore, externalities occur, which means that the social costs of economic activities, for example drainage of a wetland or water withdrawl leading to a loss of habitats, are higher than the private costs.

Another example is over-exploitation of fish stocks which often occurs when free access to the resource and the absence of adequate price signals coincide. Market failures often occur because of ill-defined, disputed or non-existent property rights and a lack of markets for goods and services provided by ecosystems. In some cases co-operation between users might take place in order to reduce some of these externalities.

Water User Associations of irrigators sharing an aquifer might serve as an example. But voluntary co-operation is rather unlikely when there is a strong heterogeneity and asymmetry of uses and users. For example, asymmetrical conflicts might occur between upstream and downstream users, between agricultural and urban users or between water withdrawal and in-stream flow. Against this background economic incentives serve as a tool for decision-making to integrate all benefits of freshwater ecosystems and all costs of loosing their goods and services.

The identification of perverse incentives and alteration to mitigate adverse effects on freshwater biodiversity might be a good starting point for action. In the European Union and in the US, for example, reforms of agricultural policy to enforce environmental concerns are being implemented: The main changes are reduction in price support measures and the removal or reduction of subsidies for the drainage of wetlands in conjunction with compensation payments to farmers for extensification and set-aside schemes and environmentally friendly agricultural practices.

In terms of incentive approaches, most European Countries and the US use a mix of regulatory instruments and subsidies rather than taxes or surcharges to change farmers behaviour. In theory, the removal of perverse incentives is one of the most cost-effective instruments for promoting conservation or sustainable use of freshwater biodiversity. Yet, perverse incentives are sometimes hard to identify without precise knowledge of the linkages between governmental activities, economic activities and ecological impacts.

Further, countries with lower subsidy levels, for example in agricultural policy, are not able to implement a similar policy switch. In addition, in many developing countries, the elimination of perverse incentives might be impeded because they are rooted in informal or ad hoc arrangements by government, and not institutionalised as procedures. Finally, people affected by the removal of incentives regularly demand compensation for income losses. Therefore, even the removal of subsidies might be an expensive tool.

## TYPES OF INCENTIVES IN ECOSYSTEM MANAGEMENT

- *Disincentives*: Removal of perverse incentives that discourage people to protect freshwater ecosystems.
- *Direct incentives*: Mechanisms which are targeted to specific objectives in ecosystem management and encourage people.
- *Indirect incentives*: Creation of an enabling environment.

Further perverse incentives occur when water prices do not reflect effects of water use on freshwater ecosystems. In this case, consumers do not attempt to improve the efficiency of use because water prices do not reflect the full economic costs of water provision, let alone the full economic value of water resources. Therefore, reforms towards full-cost pricing are an important tool in improving conditions for ecosystem management, and many countries are going this way (e.g. Mexico, Brazil, South- Africa).

In South-Africa, water charges will be increased and extended to all irrigation systems to cover not only operation, maintenance and capital costs but will also include a research levy and water conservation/ management fee. But in practice, there are numerous obstacles: in some countries freshwater ecosystem's goods are viewed as "gifts of god" which should be provided free to the people. In many countries, state-run utilities and government-regulated franchises often subsidise the provision of freshwater goods because governments are also concerned with unemployment, sector interests, getting re-elected and so on.

Direct incentives are targeted to specific objectives and include inter alia taxes, charges or levies, or grants for conservation or restoration measures. Many countries use abstraction charges or effluent charges to recover costs, or to promote environmental protection. But targeting a charge to reflect some of the determinants of impacts on freshwater ecosystems (such as seasonality and flow patterns) is complicated and many of the charges in use only show small effects on biodiversity protection. Increasing freshwater biodiversity requires both protection of existing ecosystems and restoration of degraded systems.

While grants might be an effective instrument to stop land conversion, the creation of additional wetland reserves or other restoration measures requires funds for the purchase of sites and their management. While such measures are costly in investment and enforcement, their implementation is not self-evident in political terms, in particular if compensation from the beneficiaries of enhanced biodiversity conservation to the protectors does not take place. However, practical experience suggest, for example, that incentives to private landowner might help to protect critical watersheds through the expansion of private reserves, as in Colombia.

Morocco has successfully used a revolving fund for providing loans to urban users both for water meter installation and for retrofitting water appliances, albeit the effects on freshwater biodiversity are rather indirect. Further, compensation measures are especially important if globally important freshwater ecosystems are concerned. Therefore, bilateral and multilateral transfers such as those suggested by Article of the CBD and the Global Environment Facility assist in providing necessary financial resources, although there is still room for improvement.

Indirect incentives are those which encourage people to conserve ecosystems by setting in place general enabling conditions. For example, establishing property rights might help to address failures in institutions and policies which govern natural resources. The underlying reasoning is that people, if they have no secure rights over land or nature, do not have to bear the on-site implications of degradation. Various forms of property rights can be used and have been applied as incentives for conservation, in particular in the context of rural sustainable development, such as transferring the outright ownership of land or resources to communities through the allocation of leases or concessions.

The importance of assigning property rights and of tradable property rights is frequently stressed, in particular by economists. Tradable water abstraction rights, transferable fishing rights and tradable emission rights can help to encourage conservation and sustainable use of biodiversity. Here, tradability is the crucial feature because it allows trading between users or owners in order to enhance flexibility and efficiency in use. Tradable water rights are applied in some countries.

While the US, Australia, Chile, Mexico and Brazil have introduced formal water markets, in other countries, in particular in South-East Asia informal water markets have developed. In theory tradable water rights might reduce the need for additional capital expenditures, expand the use of existing water supplies, and create an incentive to conserve by making conserved water transferable. Nevertheless, success of tradable water rights in environmental terms has been mixed.

## Acquisition of In-Stream Flow Rights

Advocates of water markets in the US state that in heavily appropriated streams, acquisitions of water rights to support in-stream flows may present the only effective option for protecting flows. Transfers allow flow restoration through voluntary action. Private organisations or state and federal agencies interested in protecting in-stream flows can work co-operatively with water right holders who are willing to transfer their water rights to in-stream use.

Water right holders are fully compensated for selling or leasing their water rights. Total acquisition of water by lease, purchase, and donation is on an upward trend. Since 1990, more than 2.3 million acre-feet of water have been acquired for in-stream use. The quantity of water traded increased considerably in 1995, with more than 600,000 acre-feet of water acquired in that year. This rise marked the beginning of an aggressive campaign by the Bureau of Reclamation to lease water for endangered salmon species in the Columbia River Basin. While there are some limitations in practice, it can nonetheless provide an important contribution for the preservation of habitats or forests depending on instream flows.

While efficiency gains are obvious, the transfer of water rights involves changes in the use of the freshwater ecosystem with sometimes negative third-party effects and environmental results. In other words, improvement in efficiency does not automatically coincide with enhanced protection of freshwater ecosystems. Therefore, effective measures to reduce such externalities stemming from water right's transfer have to be established. However, the necessary institutional or regulatory arrangements do not exist in all countries.

In some cases even negative social impacts might outweigh economic efficiency gains. Some countries have introduced the possibility for administrative bodies or NGO's to acquire "in-stream flow rights", which maintain levels in a natural stream or water body for the protection of the watershed or other ecosystems. Chile, where numerous problems have occurred in the system of tradable water rights, has in the meantime specified the minimum in-stream flow for ecological purposes.

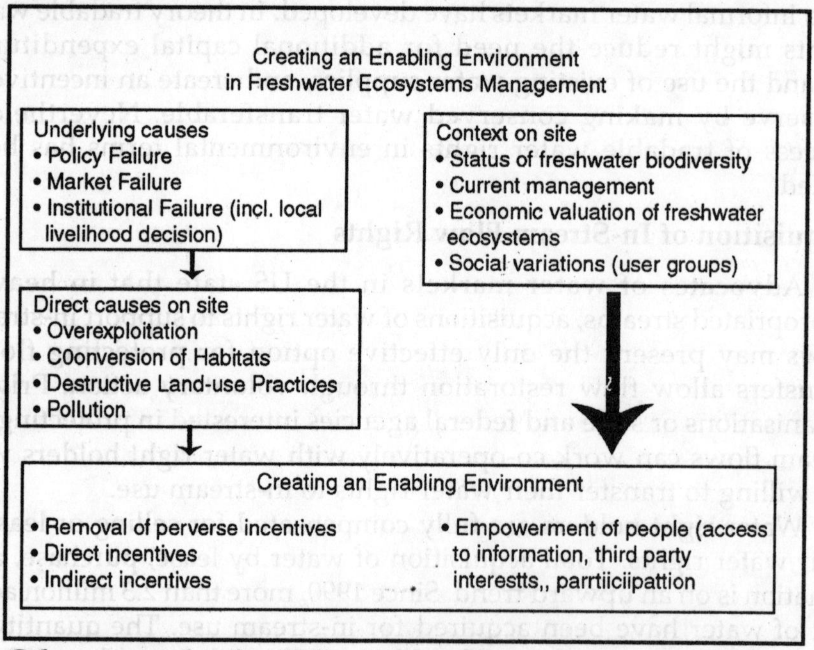


Other examples for indirect incentives are transferable fishing quota schemes, which is successfully applied in various countries, or the effluent trading scheme in the Tar-Pamlico Basin in North Carolina which helps to achieve nutrient reduction goals and addresses non-point nutrient loading while at the same time allowing municipal dischargers to trade nutrient reduction goals. Tradable water entitlements or other tradable rights are fascinating from a theoretical point of view because of both the economic efficiency gains and the potential positive environmental effects. Nevertheless, their success depends heavily on an adequate regulatory and institutional framework as well as enforcement of the rights and monitoring of their use.

Lack of incentives in conjunction with political, market and institutional failures are the most important restrictions to an effective ecosystem management. But there are some encouraging examples which demonstrate how obstacles might be overcome. However, ecosystem management is a long-term strategy in water management and can not be achieved overnight. Implementation of an ecosystem approach should follow a step-by-step approach:

- First, identification of underlying and case-specific pressures is crucial and a kind of economic valuation in conjunction with a deep consideration of social circumstances (local livelihood decisions) heavily important.
- Second, many advocates of ecosystem management oversee institutional obstacles. Therefore, institution building and capacity building must be an integral part of ecosystem management strategies. While accomplishment of the tasks of ecosystem management are already costly and difficult in "simple" cases, the accomplishment in the context of natural and institutional system complexity is sometimes impossible, especially in the short term. Therefore, ecosystem management should not insist on unrealistic institutional options with a "vision" of full co-ordination of all users and uses, but start in the framework of existing institutions and use a piecemeal approach.
- Third, in order to make ecosystem management a success story there is a clear need for the creation of an enabling environment that was traditionally neglected in classical nature conservation policy. Different incentive measures might be useful, but there concrete applicability and success depends on various institutional and economic factors, and an adequate regulatory framework must be in place. In general, incentive measures should be the outcome of a deep analysis of the nature of the problem. If ecosystem managers do not precisely know why and how people's activities result in degradation of freshwater ecosystems incentive measures can not be designed nor implemented.
- Fourth, ecosystem management is further complicated by the need for international co-operation. As success stories have

proved holistic ecosystem management is highly demanding in political terms and surly impossible if "basic" arrangements between riparian states are not in force. Therefore starting with agreements on some essentials for the preservation of freshwater ecosystems (water quality, in-stream flows) might be appropriate. Even in that cases which are often cited as success stories (Great Lakes, River Rhine etc.) ecosystem management was not the starting point but has developed on the basis of co-operation and mutual trust emerging from successful co-operation concerning realistic and short-term political goals (e.g. water pollution and/or the distribution of water quantity).

## CHALLENGES

This report has introduced a number of aspects in ecosystem management of freshwater resources: international conventions, principles of ecosystem management, and restrictions and potentials for their implementation in water management. In short, what is required for implementation is the evolution of institutions and incentives that promote an ecosystem-based approach in water management. This is not a total new insight as the need for institution building in water management is recognised since many years.

However, loss of biodiversity and ecosystem water management are relatively new tasks and experience is still limited. Further, advocates of the ecosystem management often argue from the viewpoint of natural sciences and do not sufficiently ponder on necessary institutional, organisational, instrumental or financial prerequisites. Briefly, the "who and how" of the implementation of ecosystem management is still a question that is not fully explored.

### Promote Ecosystem Management in International Agreements

Today, agreements between riparian states are confined to a limited scope, and diverging interests of states constitute main obstacles for integrated management of transboundary rivers. Riparian States already might consider minimum in-stream flow when negotiating water allocations, and water quality issues. Ecosystem management will only succeed if States increasingly recognise the usefulness of the concept. Therefore, there is a need for demonstrating benefits. Those economic and social benefits regularly exist, but in many cases ecosystem managers arguments stem more from natural science and ethics.

Further, one basic problem is that cost and benefits of the

preservation of ecosystems are not equally distributed between riparian states. Consequently, mechanism to compensate potential losers need to be establish, i.e. financial transfers or measures that generate a benefit flow in reverse direction. In addition, ecosystem management might be used as guideline while in many cases concrete action should aim to resolve less "holistic" problems (river flow, pollution). Further, there are some concrete instruments to develop and to implement:

The obligation for a strategic environmental assessment including transboundary biodiversity effects, for example, can serve to enhance knowledge and to constitute a basis for negotiations. Further, the ratification of the Watercourse Convention would be a sign of "good will" and could make negotiations easier. However, since progress on international or regional level and domestic policy reforms and institution building are two sides of the coin, ecosystem managers always may take into account national capacities (e.g. legal, organisational)

**Identify Threats and Economic Benefits of Preservation**

Much more has to be done to enhance our knowledge on status and trends of freshwater biodiversity, especially on the national and sub-national level. Ecological understanding is still weak and economic valuation of costs and benefits of measures aiming to conserve biodiversity are not frequently applied.

Indicators that could be used to measure status and trends of river basins' biodiversity are needed. The role of water experts is to co-operate in the development of indicators which can be easily applied in decision-making.

Further, elaboration and dissemination of case studies which demonstrate the usefulness of the ecosystem-based approach is a task for water experts.

**Consider Ecosystems and Social Needs**

Ecosystem management will not succeed if key goals of national and international water policy are not taken into account. Recently, for example, international water policy pursue important social goals as it aims to significantly enhance people's access to clean water. Therefore, priorities in water management frequently are not preservation of habitats, for example, but water projects that aim to meet elementary needs. Against this background, the development of win-win options on the level of programs, plans and projects is a major task and should also include effective mitigation and compensation measures.

### Find Partners for Ecosystem Management

Implementation of ecosystem management is not a self-evident process in political terms. One of the best ways to promote collaboration is the creation of networks and partnerships that links sectoral interests and stakeholders.

There are different options to succeed in the creation of partnerships for ecosystem management. For example, local communities regularly benefit from ecosystem management. Further, participation of civil society and NGOs is crucial in order to get an enhanced representation of all uses of water in decision-making.

There are even organisational options to create a forum for all uses and users (e.g. river parliaments). However, one important way to find partners is the establishment of incentive measures aiming to encourage people in preservation of freshwater ecosystems. Although experience demonstrates the benefits of incentive measures, there still is no deep knowledge on effectiveness and efficiency in terms of ecosystem management.

### Balance all Related Decision-making Levels

Decentralisation is a necessary prerequisite for ecosystem management, and the level of river basins the most appropriate. However, there still remains responsibilities on the national level, even if implementation should occur on the level of river basins. As success stories has proven, effective ecosystem management also depends on (legal and economic) decisions that has to be taken at the central governmental level.

Equally, devolution and participation of local communities are needed. Therefore, there is a kind of a "centralisation/decentralisation dilemma" in ecosystem management which only can be tackled by sufficiently flexible institutions. However, experience is rather limited.

# Chapter 4

# Tourism and Biodiversity

## THE INTERNATIONAL TOURISM INDUSTRY: OPPORTUNITIES AND THREATS FOR BIODIVERSITY CONSERVATION

## THE GROWTH OF THE INTERNATIONAL TOURISM INDUSTRY

Since the Second World War, the growth of international tourism has been phenomenal. Annual tourist arrivals worldwide increased from 25 million in 1950 to 450 million in 1990. Between 1969 and 1979, the World Bank encouraged developing countries to invest in tourism as a strategy for attracting foreign investment, and the governments of developing countries began to see tourism as a means to redistribute resources from North to South.

In the words of the World Tourism Organization (WTO), tourism became "one of the most important economic, social, cultural and political phenomena of the twentieth century". Today tourism is often described as the world's "biggest" industry on the basis of its contribution to global gross domestic product (GDP), the number of jobs it generates, and the number of clients it serves.

However, these conclusions are based largely on arrivals statistics, which focus on international tourism and therefore hide the significance of domestic tourism. These statistics may also underestimate regional tourists traveling by land rather than air or sea. The WTO estimates that the ratio of domestic to international tourism is as high as 10:1—although this varies hugely from country to country. The size of the industry and its rate of growth present both opportunities and threats for bio-diversity conservation.

### The World's Biggest Industry

Statistics produced by the World Travel and Tourism Council (WTTC) indicate that tourism generates 11 percent of global GDP, employs 200 million people, and transports nearly 700 million international travelers per year—a figure that is expected to double by 2020. According to the World Tourism Organization, international tourism:

- Accounts for 36 percent of trade in commercial services in advanced economies and 66 percent in developing economies.
- Constitutes 3–10 percent of GDP in advanced economies and up to 40 percent in developing economies.
- Generated US$464 billion in tourism receipts in 2001.
- Is one of the top five exports for 83 percent of countries and the main source of foreign currency for at least 38 percent of countries.

## Biodiversity Considerations

The Convention on Biological Diversity (CBD) defines biodiversity as "the variability among living organisms from all sources including, inter alia, terrestrial, marine and other aquatic ecosystems and the ecological complexes of which they are part; this includes diversity within species, between species and of ecosystems". The common understanding of the term "biodiversity" is all the living things on Earth and the ecological processes associated with them.

Vermeulen and Koziell note that, as such, biodiversity can be, and is, used as a synonym for "nature" or "life on Earth." Often lost in discussions of biodiversity is the emphasis on the variability and variety within species, among species, and among the ecological processes, and the key benefits these bring to humans in the form of choice—both in the present and in the future.

Biodiversity is essential to human development because of the goods and services it provides. Components of biodiversity may be used directly as food, medicine, building materials, and so on. Biodiversity provides more indirect benefits, in the form of environmental regulation, soil conservation, and pollution control. It also has what economists refer to as "non-use values"—for example, the simple enjoyment or "existence value" of some aspects of biodiversity and the option to use biological resources in the future. Many of the services biodiversity provides are not widely recognized or are not appropriately valued in economic terms. However, the combined economic value of 17 ecosystem services has been estimated at US$16.54 trillion per year.

Furthermore, one of the Working Groups for the World Summit on Sustainable Development (WSSD) reported that an estimated 40 percent of the global economy is based on biological products and processes. UNEP's Global Environmental Outlook (GEO) report on the state of the global environment highlights that, on a global scale, biodiversity is being lost at a rate many times higher than that of natural extinction. This loss is due to land conversion, climate change, pollution, unsustainable harvesting of natural resources, and the introduction of invasive species.

Human population growth, together with unsustainable patterns of consumption, increasing production of waste and pollutants, urban development, and international conflict, further contributes to biodiversity loss. Over the past three decades, major losses of virtually every kind of natural habitat have occurred, and the decline and extinction of species have emerged as major environmental issues. Although insufficient information is available to determine precisely how many species have become extinct in the past 3 decades, about 24 percent (1,130) of mammals and 12 per-cent (1,183) of bird species are currently regarded as globally threatened. So great is the concern over the rate of decline, and its implications for human welfare, that biodiversity was identified as onĕ of the five priority areas for the 2002 World Summit on Sustainable Development (WSSD).

A common strategy for biodiversity conservation has been identify priority areas and focus conservation efforts on those areas. The international system of national and regional protected areas is a clear example of this approach, and it is also implemented through a variety of international agreements: the Ramsar Convention produces a list of Wetlands of International Importance, and the World Heritage Convention identifies sites of natural heritage considered to be of outstanding value. Several inter-national conservation organizations have also adopted this strategy: BirdLife International designates Important Bird Areas (IBAs), based, inter alia, on the presence of globally threatened or endemic species; and World Wildlife Fund (WWF) has defined a "Global 200" set of priority conservation areas, with the central concept being to conserve the broadest variety of the world's habitats and the most endangered wildlife.

## Conservation International's Priority-Setting Mechanisms: Biodiversity Hotspots, Wilderness Areas, and Coral Reef Hotspots

CI has developed a priority-setting strategy that focuses its attention on biodiversity hotspots around the world. CI notes that two factors are considered for hotspot designation: "Hotspots are regions that harbor a great diversity of endemic species and, at the same time, have been significantly impacted and altered by human activities". Plant diversity is the biological basis for hotspot designation—to qualify as a biodiversity hotspot, a region must support at least 1,500 endemic plant species (0.5 percent of the global total). Existing primary vegetation is the basis for assessing human impact in a region, and a hotspot must have lost 70 percent or more of its original habitat. Overall, the hotspots have lost nearly 90 percent of their original natural vegetation.

The biodiversity hotspots contain 44 percent of all known endemic plant species and 35 per-cent of all known endemic species of birds, mammals, reptiles, and amphibians in only 1.4 percent of the planet's land area. Given the great concentration of plant and animal species in such a small and highly threatened terrestrial fragment of the world, it is extremely important that these areas receive very special conservation attention, along with research and monitoring to prevent further extinctions. The biodiversity hotspots span countries of different sizes, economic and resource endowments, and social contexts. Mass tourism, as well as nature-based and adventure tourism, is a significant revenue generator in many of these countries.

CI has also identified 10 coral reef hotspots. Eight of the 10 are adjacent to terrestrial biodiversity hotspots. Extending terrestrial conservation efforts seaward in those places offers an effective and affordable strategy for protecting global biodiversity. An additional and complementary terrestrial prioritization category used by CI is that of wilderness areas. Three major tropical wilderness areas are shown in Map —Amazonia, the Congo Forest of Central Africa, and the island of New Guinea.

They are at least 70 percent intact and are generally under less pressure from encroaching human populations than are the biodiversity hotspots, having fewer than five people per square kilometre. As such, these areas are among the last places where indigenous people can maintain traditional lifestyles. These wilder-ness areas are among the largest remaining tracts of pristine land on Earth but, compared to similarly intact desert, arctic, or boreal regions, they hold a high proportion of the planet's biodiversity. They are also of crucial importance to climate regulation and watershed protection.

### Tourism Development and Biodiversity Conservation: Linkages and Disconnects

The travel and tourism industry claims that it is well placed to con-tribute to sustainable development on the grounds that it:

- Has less impact on the environment than many other industries.
- Is based on an enjoyment of the natural and cultural

environment and so is motivated to protect them.

- Can play a positive role in awareness raising and consumer education through its vast distribution channels.
- Provides an economic incentive to protect habitat that might otherwise be converted to less environmentally friendly land uses.

The above points can be made equally in relation to tourism's potential contribution to biodiversity conservation, because biodiversity is a critical component of the natural environment that tourists enjoy. It is true, as this research project illustrates, that tourism has been growing and increasing particularly in biodiversity hotspots in the South. Given the rapid growth in nature and adventure travel within the global tourism industry during the past 2 decades, it is reasonable to assume that tourism's growth in these high biodiversity areas is linked to their relatively unique natural environments.

Destroying the environment on which the success of the industry is based is therefore like killing the goose that lays the golden egg. Tourism can, and sometimes does, make significant contributions to protected-area systems of conservation. Direct benefits from tourism to conservation can be clustered in five areas:

- A source of financing for biodiversity conservation, especially in legally protected areas.
- Economic justification for protected areas.
- Economic alternatives for local people to reduce overexploitation of wildland and wildlife resources on protected areas.
- Constituency-building, which promotes biodiversity conservation.
- An impetus for private biodiversity conservation efforts.

In the South, tourism is often the overriding justification for governments to support the creation of new protected areas. In addition, since the mid-1980s, the trend toward wildlife needing to "pay its way" and for local communities to be actively involved in conservation efforts has led to the emergence of ecotourism as a more responsible form of nature-based travel that promotes biodiversity conservation and also brings benefits to local communities.

During the seventh session of the United Nations Commission on Sustainable Development (CSD) in 1999, UNEP reemphasized the growing recognition that "the involvement of local communities in tourism development and operation appears to be one important condition for the conservation and sustainable use of biodiversity." Obligations of donors and governments under the CBD, with its emphasis on sustainable use and benefit sharing, have served to reinforce this trend, resulting in the CBD Guidelines on Sustainable Tourism in Vulnerable Ecosystems, approved in the convention's Scientific and Technical advisory body in March 2003.

**Ecotourism—Linking Tourism and Biodiversity Conservation**

"Ecotourism is an idea, a concept that is challenging tourism as we have known it. Defined most succinctly as 'responsible travel to natural areas, that conserves the environment and sustains the well being of local people,' ecotourism fundamentally reshapes the basic precepts behind tourism, which is quite simply travel undertaken for pleasure. Nature tourism, which is frequently but erroneously considered the same as ecotourism, is defined as travel to unspoiled places to experience and enjoy nature.

Its close cousin, adventure tourism, is described as nature tourism with a kick—nature tourism with a degree of risk taking and physical endurance. Nature and adventure tourism focus on what the tourist is seeking. In contrast, ecotourism is qualitatively different. It focuses on what the traveller does, plus the impact of this travel on both the environment and the people in the host country. Ecotourism posits that this impact should be positive. Ecotourism is not, therefore, simply another niche market within the tourism industry. Rather, ecotourism is a philosophy, a set of practices and principles that, if properly understood and implemented, will transform the way we travel."

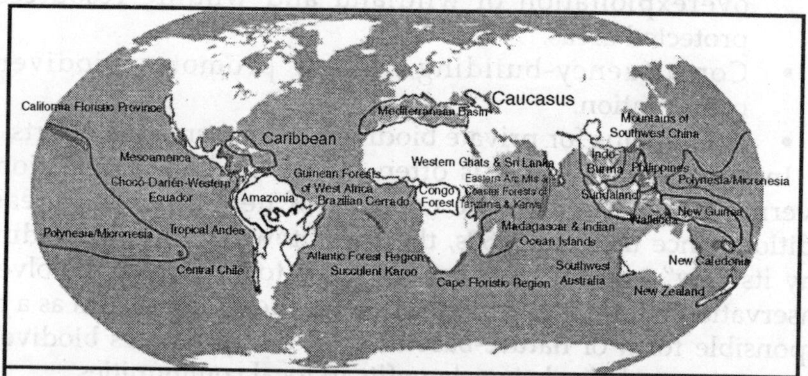


**Fig.** Hotspot, Major Tropical Wilderness Areas and Coral Reef Hotspots.

Communities that receive significant income from tourism may be motivated to conserve bio-diversity. However, if benefits are small—

or not sufficiently linked with conservation inputs—they may be reinvested in activities that undermine biodiversity conservation such as livestock rearing.

Sustainable tourism has emerged as a more responsible form of mass tourism development. In the past, traditional mass tourism developments have been a major threat to biodiversity conservation because management controls and effective planning mechanisms have been lacking. Drawing from the concepts of ecotourism, namely that tourism should "conserve the environment and sustain the well-being of local people", sustain-able tourism seeks to minimize the negative footprint of tourism developments and at the same time contribute to conservation and community development in the areas being developed.

It might be expected that tourism development following the principles associated with ecotourism would go hand in hand with biodiversity conservation and improvements in rural livelihoods. In many instances, tourism has been instrumental in delivering scarce funds for conservation and providing local people with an economic incentive to protect biodiversity from other, potentially more damaging forms of development such as mining, logging, or consumptive use of wildlife. Maintaining an attractive resource base has in turn continued to attract more tourists and sup-port a healthy tourism industry, thus generating more funds for conservation.

A mutually supporting circle of success can develop. However, this positive relationship is not always the case, particularly where tourism development occurs without management standards and guidelines that seek to pro-mote conservation of nature and deliver tangible benefits to local communities. Moreover, whatever the form of tourism developed, infrastructure and facilities for the tourism industry usually require significant tracts of land and building materials. Tourism development regularly takes place in a rapid and unplanned manner, resulting in total landscape transformation in a very short period of time, often leading to deforestation and drainage of wetlands. Such habitat disruption can result in significant loss of biodiversity. The problem is exacerbated by the fact that a lot of tourism occurs in fragile areas (e.g., coastal zones, mountains, protected areas) or areas of high biodiversity.

## Sustainable Tourism and Biodiversity Conservation

Sustainable tourism (which draws on the principles of ecotourism) can directly contribute to biodiversity conservation by:

- Offering less destructive livelihood alternatives to local communities and landowners in buffer zones and conservation corridors, away from unregulated logging, intensive cattle-ranching, monoculture, hunting, and unsustainable tourism.

- Providing an incentive for public and private landowners in critical ecosystems to permanently conserve biodiversity-rich properties, by offering revenue-producing, low-impact economic use.
- Providing protected-area managers with additional financial resources from visitation and donations.
- Raising visitor awareness, promoting community involvement and interest in conservation issues, and generating political support for conservation through environmental education during travel.

The very nature of the main-stay product of the tourism industry—sand, sea, and sun—high-lights the fact that the industry is very dependent on coastal areas. Alongside increasing urbanization and industrialization, UNEP's Global Environment Outlook (GEO) report notes that uncontrolled, mass tourism is one of the root causes behind coastal degradation today. As well as its direct environmental impacts, resource depletion can also have indirect socioeconomic effects, as essential resources become scarce for local people.

The World Conservation Union (IUCN) notes that this can, in turn, have additional negative impacts on biodiversity, by concentrating local resource use in smaller areas and/or by undermining local resource management systems. In addition to resource depletion and habitat disruption, littering and water pollution are problems associated with mainstream tourism that can have negative consequences for biodiversity conservation. The littering problem is exacerbated in remote areas, where waste collection can be logistically difficult (e.g., on mountains, in the middle of the ocean).

Waste disposal from cruise ships has been problematic, as they have limited capacity to carry all their waste until they reach their home port, and destination ports have limited incentive (and capacity) to accommodate periodic discharges. However, "most of the major cruise lines have begun to implement comprehensive waste management programs and wastes such as glass, cardboard, aluminum and steel cans are processed onboard through crushing, reuse and/or recycling and incineration".

Construction of hotels, recreation, and other facilities often leads to increased sewage pollution. Wastewater has polluted seas and lakes surrounding tourist attractions, damaging the flora and fauna. Coral reefs are at a particular risk from unplanned tourism development. Holden notes that, as well as being mined for building materials, reefs suffer from sewage runoff that stimulates the growth of algae, covering the filter-feeding corals and hindering their ability to survive. In addition, reefs are often damaged by the activities of careless tourists—as divers and snorkelers kick and stand on coral, for example, or boats and jet skis scrape the surface of the reef.

Furthermore, dive/snorkel boat operators may throw their anchors into corals, and local entrepreneurs often break off pieces of coral to sell as souvenirs. Mountains are also popular locations for tourism and, because of their fragile soils, they are particularly sensitive to environmental impacts. Deforestation from tour-ism (through construction, use of wood for fires, etc.) can have direct implications for habitat conservation and watershed management and can also increase the likelihood of landslides.

It is these negative impacts of unmanaged tourism development on the environment and local cultures that gave rise to ecotourism in the 1980s and 1990s as an alternative set of principles and practices to harness tourism's economic potential for biodiversity conservation and sustainable development.

During the last decade, ecotourism has emerged from small model projects demonstrating how tour-ism can be a catalyst for conserving nature and promoting the well-being of local peoples into a wider set of sustainable tourism principles that can be applied across a larger segment of the travel and tourism industry. However, just how far, and to what degree, these principles will ultimately be able to transform the mass tourism industry to be a more positive force for biodiversity conservation remains to be seen.

## Tourism's Resource Consumption

Using consumption averages from various countries, statistics from WTO, and estimates of national tourism in relation to international arrivals, UNEP proposed some estimates of the order of magnitude of resource consumption from tourism. If the global tourism industry were represented as a country, it would consume resources at the scale of a northern developed country. International and national tourists use 80 percent of Japan's yearly primary energy supply (5,000 million kWh/year), produce the same amount of solid waste as France (35 million tons per year), and consume three times the amount of fresh water contained in Lake Superior, between Canada and the United States, in a year (10 million cubic meters).

# MAPPING TOURISM'S GLOBAL FOOTPRINT: IMPACTS ON BIODIVERSITY AND LOCAL LIVELIHOODS

## THE MAPS

Earlier Chapter has highlighted the broad relationships between tour-ism development, biodiversity, and local livelihoods. This chapter reviews this relationship in more depth, focusing on the impact of tourism in the biodiversity hotspot countries. The maps presented serve to illustrate the following key issues:

- Although most biodiversity is concentrated in the South, many major tourism destinations in the North (e.g., the Mediterranean, the California Coast, Florida Keys) also coincide with biodiversity hotspots.
- Although they receive fewer tourists overall than the North, many economically poor, but biodiversity-rich, countries in the South receive large numbers of tourists.
- Many hotspot countries in the South are experiencing very rapid tourism growth.
- Over one-half of the poorest 15 countries fall within the biodiversity hotspots, and in all of these, tourism is already significant or is forecast to increase.
- In many destinations within hotspot developing countries, biodiversity is the major tourism attraction.
- Forecasts suggest that tourism will become increasingly important in hotspot countries—particularly in Southeast Asia—and this will require careful planning to avoid negative impacts on biodiversity.

The maps can also be used to illustrate the potential impacts of tourism in different countries or regions—for example, plotting the number of tourism arrivals against the population of each country allows us to predict where tourism pressure is likely to be high and environmental and social impacts more severe.

## Is Tourism Significant in Biodiversity Hotspot Countries?

The hotspots map shows that, on a global level, the majority of hotspots are concentrated in the South. A map of international tourist arrivals by country for 2000 shows, however, that the majority of tourist arrivals are in the North: North America, Western Europe, and Russia stand out as significant areas for tourist visitation.

This finding is borne out by a map of arrivals by region, which also shows that Southeast Asia and South America receive medium levels of arrivals. Africa, South Asia, Oceania, and Central America experience lower levels of arrivals at the regional level. An analysis based solely on total arrivals figures can obscure the pressure of tourism in some cases, as it does not take into account the size of the country to which the tourists are arriving and the amount of tourism infrastructure and planning in place.

Map shows fewer tourists arriving in the Caribbean than in the United States, for example, but when the relative sizes of these regions are taken into account, the implications of even this lower number of arrivals becomes clear. Although the Caribbean region, a major biodiversity hotspot, accounts for only 4 percent of international tourist arrivals, tourism plays a major role in many Caribbean economies, accounting for 15.5 percent of total employment, or one in 6.4 jobs. The biodiversity impacts of tourism development in the Caribbean may be much more significant than the statistics initially convey.

In addition, data on the volume of tourism provide no indications as to how tourists are distributed within a particular country—they may be extremely concentrated in some areas and virtually nonexistent in others. The Caribbean is again a good example, where the vast majority of tourism impacts happen at the coast, the site of often critically endangered coral reefs. Obviously, the distribution of tourists in relation to sensitive areas will affect the impact of tourism both on local livelihoods and on biodiversity.

## Tourism is Significant and Growing in Poor, Biodiversity-rich Countries

Although they receive fewer tourists overall than the North, parts of the South receive large numbers of international arrivals, and many of these coincide with hotspots: Mexico (Mesoamerica hotspot), Brazil (Atlantic Forest and Cerrado), South Africa (Cape Floristic Region), Thailand (Indo-Burma), Malaysia, and Indonesia (Sundaland and Wallacea) stand out as countries with high levels of international tourism arrivals, particularly during the past decade. At a subnational level, the Cape Floristic Region of South Africa is an important tourism destination, and the resort island of Bali falls squarely within the Sundaland hotspot.

It is also important to note that Map shows only the numbers of international arrivals, yet domestic tourism is also highly significant in Mexico, China, South Africa, and Thailand. In Brazil, domestic tourists provided six times more room nights in classified hotels in 2001 than the 5.5 million foreign tourists. Ghimire notes that in Mexico, it was estimated that as much as 40 percent of the country's population participated in domestic tourism activities in 1994.

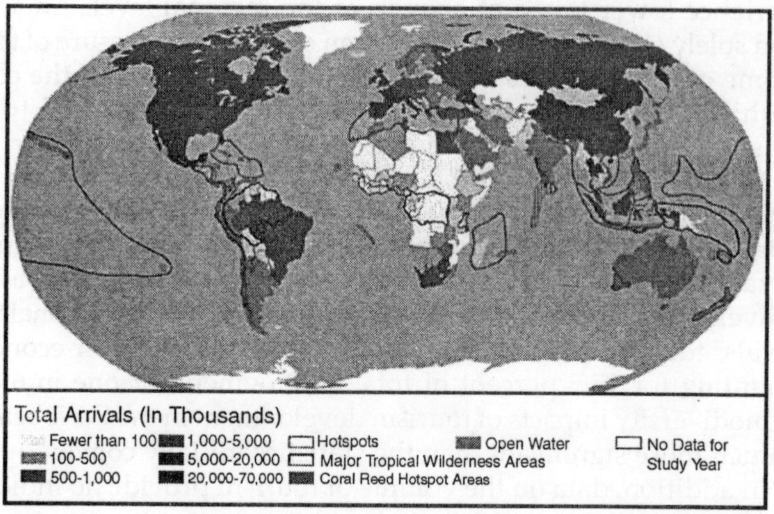


**Fig.** International arrivals Circa 2000.

In Thailand, domestic tourists outnumber international tourists at all major attractions. On the Philippine island of Palawan, listed by CI as one of the world's most threatened biodiversity hotspots, domestic tourism accounts for more than 50 percent of arrivals in 2003. Although the international arrivals represented in figures are not exclusively vacation tourists, tourists can be considered a large, if not the largest, segment of those arrivals; the statistics therefore do represent significant increasing tourism travel to each biodiversity hotspot country.

Furthermore, although North America may receive many arrivals, Map shows that the average annual growth rate over the last 10 years has been slower in North America than in other biodiversity hotspot countries such as Brazil. South Africa is among several hotspot countries where the number of international arrivals is not only large but also rapidly growing. Tourism in Laos and Cambodia (Indo-Burma) has also grown dramatically, as it has in Vietnam and Burma.

Of particular importance, Table identifies 22 hotspot countries where visitor arrivals have increased by more than 100 percent between 1990 and 2000. At the top of the list, Laos shows a staggering tour-ism increase of over 2,000 percent. Although starting from a small base (14,000 international arrivals in 1990), if Laos follows the pattern of its neighbour Vietnam, which has increased from 250,000 to 1,890,000 tourist arrivals in the last decade, the implications could be very significant in terms of negative impacts on biodiversity. Not only did the number of international arrivals in China top 10,000,000 in 1990, but it nearly tripled to 31,000,000 in 2000. By contrast, the United States, while capturing a larger number of international arrivals (51,000,000), has experienced only a comparatively modest growth rate of 29 percent in the last decade.

These patterns of growth are particularly important, since it is reasonable to assume that a significant percentage of new tourism facilities in developing countries high in biodiversity will be built on coastal and natural destinations harboring threatened ecosystems.

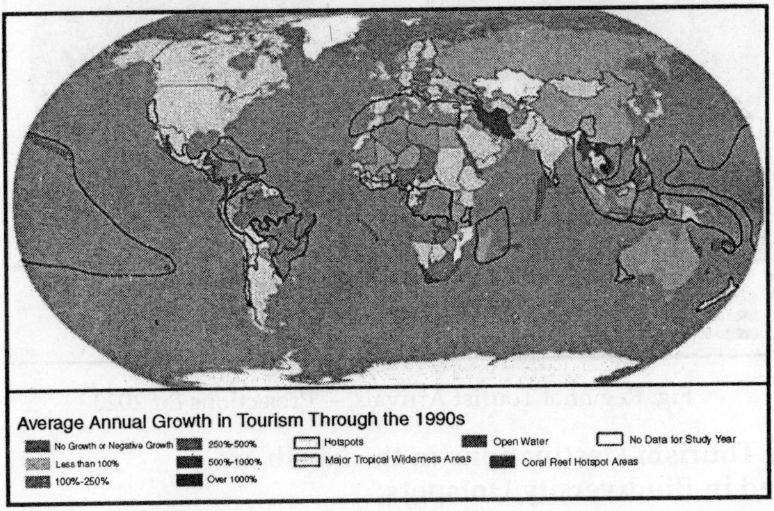


**Fig.** Average Annual Growth 1990-2000.

Further prioritizing exercises may be proposed: for example, focusing on countries with high volumes of international arrivals (38 hotspot countries show over 1 million international arrivals per year, and 16 of them show over 5 million arrivals per year); or combining arrival numbers with growth rates—12 of the 22 hotspots countries with over 100 percent growth rates had over 1 million international tourists in 2000 (Australia, Brazil, China, Costa Rica, Cuba, the Dominican Republic, Indonesia, Israel, Macao, Peru, South Africa, and Vietnam). These countries clearly need to ensure that biodiversity considerations are incorporated into tourism development strategies and policies and that tourism is considered in strategic biodiversity action plans.

Looking forward to 2020, regional forecasts prepared by the World Tourism Organization suggest that tourism will become increasingly important in hotspot countries. South America, south-ern Africa, and Oceania are all expected to experience significant growth in numbers of tourists, but the Southeast Asia region stands out as one where the increase is likely to be particularly dramatic. This projection implies that, as the home to four biodiversity hotspots and one major tropical wilderness area, this region will require very careful tourism planning if it is not to suffer a serious negative impact on biodiversity.

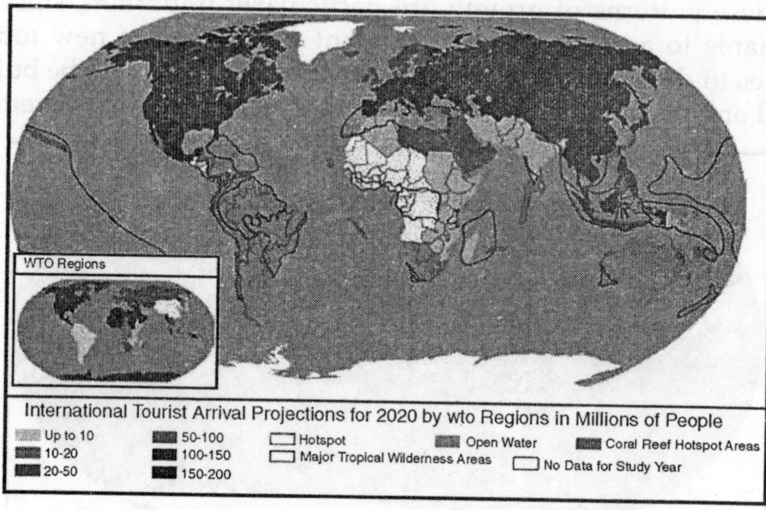


**Fig.** Regional Tourist Arrivals — Projections for 2020.

## Prime Tourism Destinations in the North are Located in Biodiversity Hotspots

Tourism in the North also has significant implications for bio-diversity conservation, because biodiversity hotspots also occur in these northern destinations: the California Floristic Province, the northern part of Mesoamerica, the Mediterranean Basin, the Caucasus, and the mountains of south-central China, for example.

From the level of analysis of the global mapping exercise, it is not possible to examine the distribution of visitor arrivals across these regions—it is theoretically possible that none of China's tourists visit the south-central mountains, for example.

But in actuality, tourism pressure is well documented in this area of China, and significant growth in China's domestic tourism is anticipated, with some major tourism development projects already under way. The Mediterranean is the most visited tourism region in the world, accounting for 30 percent of inter-national arrivals and 25 percent of receipts from international tourism.

The number of tourists in the Mediterranean countries is expected to increase from 260 mil-lion in 1990 (with 135 million to the coastal region) to 440–655 mil-lion in 2025 (with 235–355 million to the coastal region). It can also be noted that "the construction of infrastructure and the direct impacts of people using and trampling ecosystems remains a key threat to coastal areas in Turkey, Cyprus, Tunisia, Morocco, and Greece."

## Tourism, Biodiversity, and Poverty Reduction

We have already noted above that the majority of biodiversity hotspots are located in the developing countries of the South. In light of the linkages between biodiversity and tourism, and between biodiversity and sustainable livelihoods, it is clear that no biodiversity conservation strategy based on tourism alone is likely to succeed unless it incorporates some poverty reduction goals.

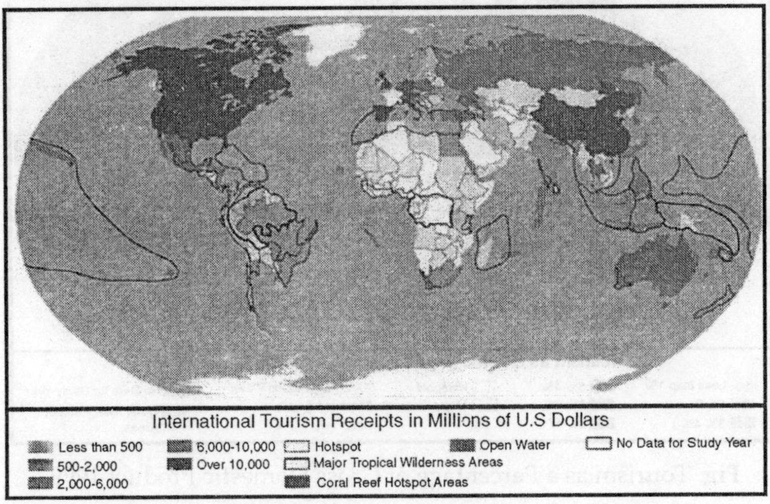


**Fig.** International Tourism Receipts Circa 2000.

Developing countries currently have only a minority share of the international tourism market (approximately 30 percent), but their share is growing rapidly. International tourism arrivals in developing countries as a group have grown by an average of 9.5 percent per year since 1990, compared with 4.6 percent worldwide. In these countries, tourism makes important contributions to the national economy through foreign exchange earnings, employment, and GDP.

On average, international tourism receipts account for around 10 percent of export revenues of developing countries. The United Nations Conference on Trade and Development (UNCTAD) notes that tourism is a principal export of 49 least-developed countries and number one for 37 of them. Tourism's contribution to GDP varies from 3 to 5 percent in Nepal and Kenya to 25 percent in Jamaica; contribution to employment is estimated at 6–7 percent in India and South Africa. Maps illustrate the significance of tourism as a percentage of GDP in developing countries.

The maps demonstrate that in the industrialized North, high levels of tourism receipts correlate to their significance in terms of GDP. In the less industrialized countries of the South, however, even low levels of tourism receipts can be very important to the national economy. In short, even modest levels of tour-ism, carefully planned and imple-mented, can be a positive force for biodiversity conservation and local economic benefit.

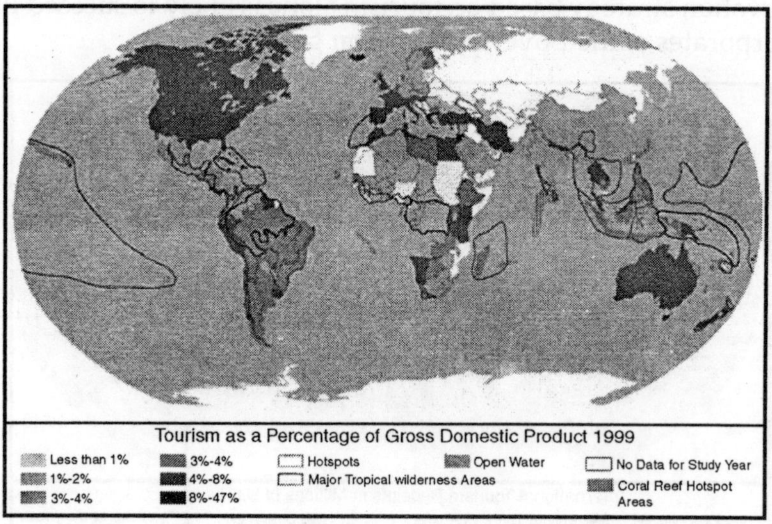


**Fig.** Tourism as a Percentage of Gross Domestic Product 1999.

Although we cannot accurately determine the degree to which tourism is directly dependent on biodi-versity, we can assume with confidence that in many hotspot coun-tries, such as Australia, Belize, Brazil, Costa Rica, Kenya, Madagascar, Mexico, South Africa, and Tanzania, a significant proportion of tourism's GDP contribution can be directly linked to attractions and destina-tions in biodiversity hotspots, where biodi-versity itself represents the primary tourism attraction.

Tourism is clearly of great economic significance to developing countries. However, that significance varies widely from country to country, with those economies most dependent on tourism tending to be small island states: The Caribbean is the most tourism-dependent region in the world, and the Maldives the most tourism-dependent country. Although these countries are not the poorest in the world (they are classified by the World Bank as middle-income on the basis of indi-cators such as numbers of people living on less than US$1/day), they still contain significant numbers of impoverished people.

Of the poor-est 100 countries, however, well over half have a tourism industry that is growing and/or significant. Table shows 6 of the world's 15 poorest coun-tries where tourism is significant or growing. All are in biodiversity hotspots. Up-to-date poverty data (from the World Bank Development Indicators) are not available to plot a comprehensive map for all biodiversity hotspots. However, an analysis of tourism arrivals against the Human Development Index (HDI) of the United Nations Development Programme (UNDP) serves to illustrate the overlap between levels of develop-ment, biodiversity, and tourism.

In particular, several hotspot countries have a low HDI rating and high levels of visitation—for example, Brazil, Indonesia, and South Africa. The map also illustrates the overlap between hotspots and countries with a low HDI rating (Cambodia, Gabon, Ghana, Guatemala, Indonesia, Ivory Coast, Madagascar, Papua New Guinea, southern Nigeria, Tanzania).

**Table. Significance of International Tourism to Poor, Biodiversity-Rich Countries**

| Country | Hotspot | Percentage of population below US 1$ a daya | Percentage contribution of tourism industry to GDPb |
|---|---|---|---|
| Nigeria | Guinean Forest | 70 | 0.5 |
| Madagascar | Madagascar and Indian Ocean Islands | 63 | 3.8 |
| India | Western Ghats and Sri Lanka | 44 | 2.5 |
| Honduras | Mesoamerica | 41 | 4.4 |
| Ghana | Guinean Forest | 39 | 5.5 |
| Nepal | Indo-Burma | 38 | 4.5 |

A key question that might be asked is, if these countries are apparently doing so well in tour-ism and are so well endowed with biodiversity, why are they still so poor? Some argue that because for-eign, private sector interests often drive tourism, it has limited poten-tial to contribute much to poverty elimination in developing coun-tries. This can apply to biodiver-sity-based tourism, as well as other forms of tourism development.

Tourism is often noted for having high levels of revenue "leakage," and of the revenue that is retained in the destination country, much is captured by rich or middle-income groups—not the poor.

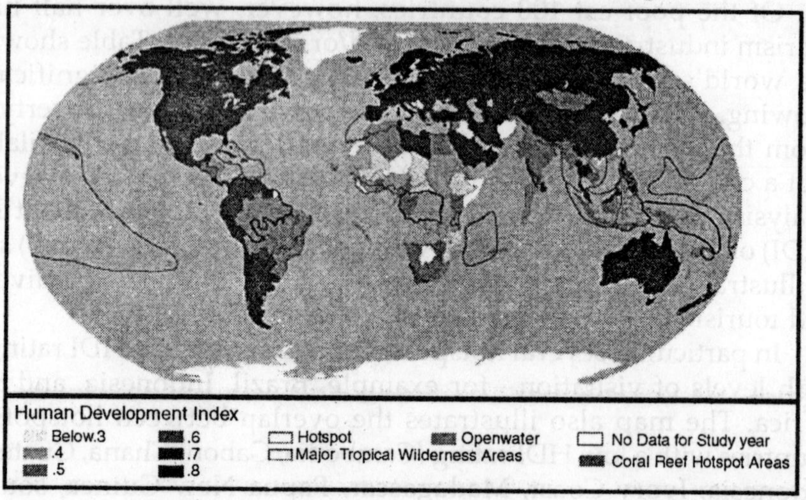


**Fig.** UNDP Human Development Index 2000.

Tourism is also a volatile industry, being extremely susceptible to events that are difficult to control—natural disasters, exchange rate fluc-tuations, and political unrest. For example the 2002 terrorist bomb-ing on the resort island of Bali led to an immediate drop in tourism arrivals and it was almost a year before tourism on Bali began to increase again. In poor countries, tourism can have a particular effect on the poor themselves, causing displacement, increased local costs, loss of access to resources, and social and cultural disruption.

Tourism does, however, have a major advantage over other forms of development (such as timber extraction, mining, etc.) with respect to biodiversity conservation and poverty reduction: Not only is tourism highly dependent on the natural and cultural environ-ment—assets that the poor have and on which they can capital-ize—but, properly managed, it can contribute to biodiversity conserva-tion, which can directly support poverty reduction. Strategies for making tourism more "pro-poor" have shown some success at the local level. Scaling these approaches up and applying them to biodiversity-based tourism could result in positive synergies between tourism growth, biodiversity conservation, and human development in the future.

## Analyzing the Maps to Assess Impacts

The ratio of visitors to local resi-dents is used by the World Tourism Organization as a core indicator of the social impact of tourism, and the map below illus-trates that this ratio can be extraor-dinarily high in some countries, with tourists outnumbering local residents in certain areas. Hotspot countries or areas that stand out are Australia, Botswana, Eastern Caribbean, New Zealand, northern Mediterranean, Malaysia, Mexico, and Uruguay.

Furthermore, tourism is likely to increase in the next 20 years, including in areas where visitor pressure is already high. It should be noted, however, that this visitor-to-resident ratio is an extremely rough measure of impact. The local distribution of the tour-ists, the activities they engage in, and the cultural differences between tourists and residents need to be considered as well, before the potential impact can be fully assessed.

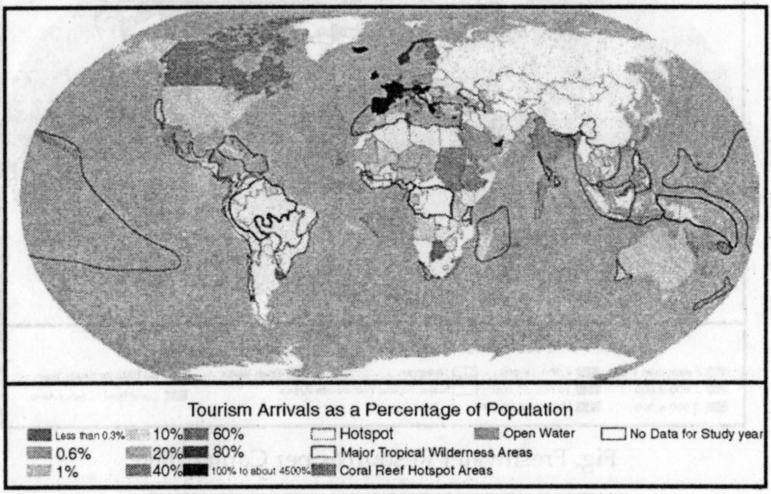


**Fig.** Arrivals as a Percentage of Population.

The ratio of tourists to local residents can be used in conjunc-tion with other data to highlight potential environmental impacts. Water use, for example, can be a serious issue with respect to tourism development. Map shows that in some countries the availability of freshwater resources is very limited, yet some of these countries have tourism intakes far higher than their total population.

Tunisia, for example, is conspicuous as a country within the Mediterranean hotspot with limited per capita freshwater resources and very high tourist-to-resident ratios. The Caribbean, Mexico, and South Africa also stand out as hotspot areas with high levels of visitation (and sustained growth in visitation over the 1990s) and low levels of available fresh water.

Water use is a particular prob-lem associated with hotels, as tourist consumption of water is often many times higher than that of the local people. This can result in water shortages and degradation of water supplies, as well as increased wastewater discharge, all of which can affect wetlands. The problem is particularly acute in hot, dry coun-tries (both in the North and South), where available resources can be in short supply, yet tourist demands on water (for swimming pools, show-ers, etc.) are high because of the climate. The vast quantities of water required to maintain golf courses (a rapidly increasing form of tour-ism in the South) is another issue of concern. An average golf course soaks up at least 525,000 gallons of water per day, which can severely affect fresh water availabil-ity in certain areas.

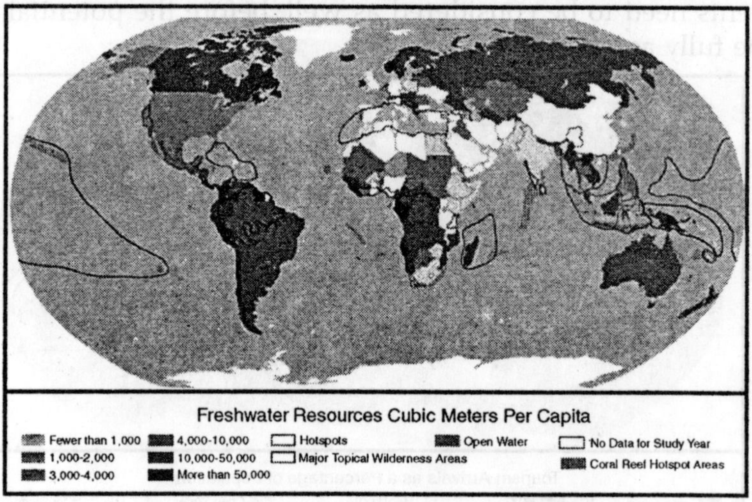


**Fig.** Freshwater Resources per Capita 2000.

**Cancun, Mexico: The Impact of Tourism Development**

Prior to its development as a tourist resort in the 1970s, only 12 families lived on the barrier island of Cancun. The entire area that now comprises the state of Quintana Roo was made up of relative-ly untouched rain forests and pristine beaches and was inhabited by an indigenous Maya population of about 45,000. Today, Cancun has more than 2.6 million visitors a year and has more than 20,000 hotel rooms, with a permanent population of more than 300,000.

Environmental and social impacts were given secondary importance in the development plan for Cancun. For instance, no provisions were made to house low-income migrants who now work and live in the area. As a result, a shantytown developed, in which the sewage of 75 percent of the population is untreated. The mangrove and inland forests were cut down, swamps and lagoons were filled, and dunes were removed. Many bird, marine, and other animal species vanished.

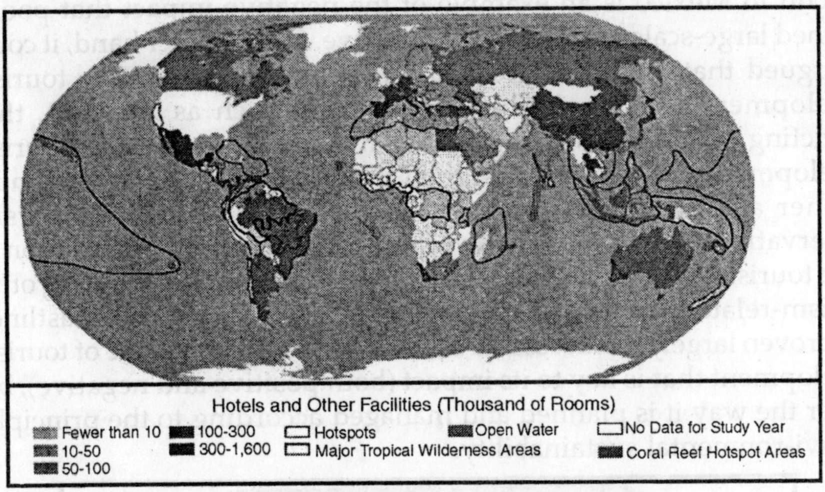


**Fig.** Hotels and Other Facilities—Rooms Circa 2000.

Map forecasts increased tour-ist arrivals in the arid countries of North Africa and the Middle East, where the tourist-to-resident ratio is already very high, and in the hotspot region of the Mediterranean. The conclusions that can be drawn from these maps are somewhat limited, since some very dry countries have a very small population (e.g., Namibia) and so do not show up as a problem on Map. However, they are highly vulnerable to an increase in water use as a result of tourism or any other extractive use and serve to highlight some of the pressures that can be associated with an increase in tourism.

Thus, the forecast map emphasizes the need for proper planning if continued growth of tourism is not to impinge even further on water availability and its relationship to biodiversity-rich wetlands and the well-being of local residents. Maps depicting the scale of hotel development and levels of occupancy, are also useful indicators of potential impacts—especially in light of the connection between hotel development and water use highlighted above. Some countries appear to have overde-veloped their hotels. Thailand, for example, shows a high level of capacity and low level of occupancy (less than 50 percent). Indonesia also shows a similar low level of occupancy and high-level capacity, although arrivals are projected to grow dramatically through to 2020.

Bearing in mind the envi-ronmental impacts associated with building and infrastructure development and the potential consequences for biodiversity con-servation, this should be a point of concern in hotspot countries, and it underscores the need for careful planning of any further development. The mass tourist resort of Cancun in Mexico is an example of the negative impact that poorly planned large-scale developments can have. On the other hand, it could be argued that, on a wider scale, it is better to concentrate tourism development into a relatively small area (such as Cancun), thus restricting, spatially, its impact. This can work even with large tourism developments if they are designed in an environmentally friendly manner and revenue from them is used to support biodiversity conservation elsewhere. However, maintaining the concentration of large tourism developments in specific destinations and avoiding other tourism-related sprawling developments, especially along coastlines, has proven largely unsuccessful. It is not necessarily the scale of tourism development that is key to its impact (both positive and negative), but rather the way it is planned and managed according to the principles of envi-ronmental sustainability.

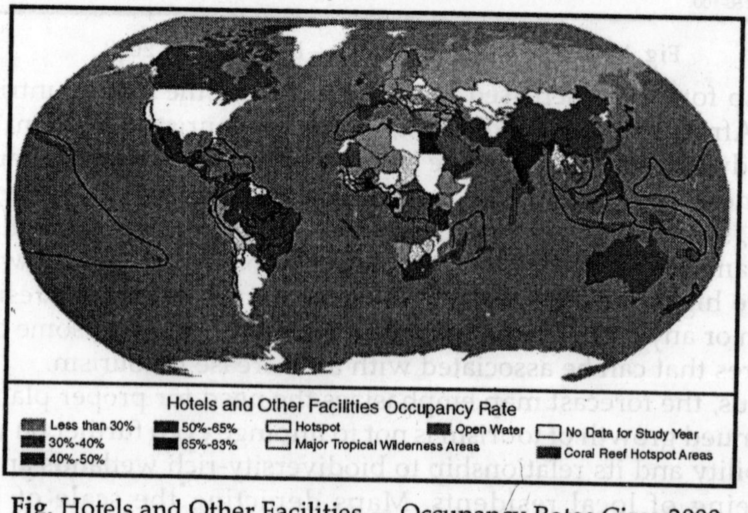


Fig. Hotels and Other Facilities — Occupancy Rates Circa 2000.

Map shows high levels of hotel occupancy in the Caribbean. While on the one hand this illustraces that infrastructure has not been developed unnecessarily, unlike some areas, it also emphasizes the high levels of tourist traffic in this region and the potential effect of related social impacts.

## KEY DECISIONMAKERS REGARDING TOURISM AND BIODIVERSITY CONSERVATION

In order to effect changes in the tourism industry toward more conservation-friendly processes in biodiversity hotspots, it is important to identify the key players whose decisions will ultimately affect biodiversity. Furthermore, recommendations to public and private planners and policymakers at various levels of decision making have to take into consideration the limits imposed by their sphere of influence and immediate objectives. Key-player analysis allows policymakers and practitioners to understand the concerns and interests of different parties and the means by which each agent contributes to tourism development and to biodiversity conservation or loss.

The tourism industry can be seen as a network of economic and political agents, processes, and resources. The interactions between these elements will ulti-mately define whether the impacts on biodiversity will be positive or negative. For this study, we have identified governments, the private sector, development agencies, and local residents as key players, given their central influence regarding the impacts of tourism on biodiversity. Other stakeholders, such as tourists, nongovernmental organizations (NGOs), intergovernmental agencies and experts, academics, and consultants, also have important roles, but their actual contribu-tion depends on their ability to influence the central players. This chapter will identify the roles and interests of various stakeholders related to conservation in the biodiversity hotspots.

## TOURISM IMPACTS BIODIVERSITY HOTSPOTS THROUGH

- Infrastructure-related development, which is primarily financed and managed at the governmental level. This includes methods of access (roads, trails, airports, and transportation); water sourcing and treatment facilities; energy production and distribution; and waste processing. Private investment in infrastructure development, and interpreta-tion and visitor-management structures, often results in landscaping and construction-related impacts on biodiversity.

- Construction of facilities directly related to tourism (accommodation and meeting structures, catering, shopping centers, marinas, and administrative facilities).
- Indirect developments from tourism, such as urban development for employee housing; secondary real estate, such as tourist homes; and urban sprawl.
- Indirect influences on economic and trade policies and strategies related to tourism development (changes in local traditional economic practices due to transition to tourism-oriented activities; changes in management practices due to globalization; changes in conservation-related investments, such as park management, and in environmental management due to financial burdens from tourism-related loans, etc.)

The different phases of tourism development have different impacts on biodiversity conservation. Decision making on siting, design, and planning have different effects than technological decisions on management and operations of tourism, including water and waste pollution, resource consumption, and supply chain management. Similarly, biodiversity is affected by choices made by tourists regard-ing their activities such as hiking, boating, and sports.

Trends in the tourism marketplace also determine effects on biodiversity. Often, an attractive tourism destination in a biodiversity-rich area may experience a pronounced growth phase in accom-modation development, ultimately leading to over-development, with serious negative impacts on biodiversity. Heavy price discounting in tourism resorts, with low-margin, high-volume vacation packages and uncontrolled competition, can lead to loss of economic feasibility.

In these cases, environmental management costs are not covered, and environmental degradation ensues, ultimately causing a crisis or even the collapse of a tourism destination, again with serious effects on local biodiversity.

Ideally a timely assessment of this negative trend will lead to a multi-stake holder process that ensures proper oversight. Tourism development can be contained and better management systems can be provided for natural resources. In addition, financial mechanisms can be set up to use a small percentage of tourism revenues for environmental management. For instance, the municipalities of Calvia, in Spain, and Rimini in Italy, both located in the Mediterranean hotspot and heavily dependent on tourism, experienced overdevelopment and environmental degradation throughout the 1970 and 1980s.

These tourism destinations faced such challenges by implementing several radical measures. In the case of Calvia, a Local Agenda 219 process led to the closing and even deliberate destruction of hotels, landscape renovation, the creation of additional protected areas, and the establishment of an environmental levy on hotel room sales, with extensive public awareness and marketing campaigns.

In Rimini, coastal eutrophica-tion of the Adriatic Sea led to algal blooms and heavy fish mortality in 1985, with ensuing odors and pol-lution causing tourism losses. The tourism industry pressured local authorities to engage agribusinesses and hotel chains to reduce use of fertilizers and improve waste and sewage management. The environmental improvements were accompanied by public awareness and marketing campaigns to improve the city's image, and visitor numbers increased.

## The Decisionmaking Process for Tourism Development

In 2001, UNEP reviewed 12 case studies of tourism resort development in various ecosystems in order to investigate how decision making affects biodiversity. On the basis of this analysis, the process can be simplified as follows:

- A group of local investors, often owning biodiversity-rich land, team up with potential resort builders and hire professional intermediaries called developers, whose role it is to bring together all resources and players that will determine the feasibility of the resort.
- The developers look for outside private investors and examine the interest of partners such as tour operators and air and cruise carriers, based on per-ceived market potentials.
- The group contacts local and national government, looking for support such as
  - Infrastructure (free land, air-ports, roads, water supply, and sewage/waste management, etc.)
  - Flexible land-use regulations (appropriate for clusters of resorts).
  - Tax breaks and incentives.
  - Soft and subsidized loans.
  - Attractive public land or parks that could be the base for tour products.

The process can also be initiated by local politicians and/or inves-tors who pressure the government to offer support and then attract outside investors. Trade associations (representing tour operators, hotel chains, and air carriers, etc.) are often partners in lobbying gov-ernment, whose driving interests are job generation and future tax revenue. In some cases, tourism development is financed by multi- and bilateral development agen-cies, under subsidized development aid loans. The terms of these loans may or may not be supportive of biodiversity conservation.

- Once funding is in place the resort is built. This can occur with or without an environmental impact assessment, depending on local reg-ulations. Unfortunately, the UNEP report points out that decisions about siting, design, technology, and product development are often made only from the perspective of corporate efficiency and customer relations; community expectations and conservation of local and regional biodiversity are not usually considered.

The sections below review the roles of the different stakeholders in influ-encing this process and ensuring that tourism is developed in a manner that supports biodiversity conser-vation and benefits local people.

These descriptions are followed by recommendations for each group of stakeholders.

## GOVERNMENTS

National governments set the frame-work for tourism development and biodiversity conservation through policy and legislation. Different government departments may be responsible for determining policy and associated instruments for tour-ism development and biodiversity conservation. Some examples of these include

- Laws and regulations defining standards for tourism facilities, access to biodiversity resources, and land-use regulation and zoning.
- Design, development, and regula-tion of supporting infrastructure (water, energy, roads, airports, etc.)
- economic instruments defined in policy, such as incentives for sus-tainable tourism investment and the creation of private reserves.
- Standards for health and safety, quality controls and regulation of business activities; these are aimed at protecting consumers and at meeting the needs of residents—including traditional communities and indigenous people—and protecting their lifestyles.
- Establishment and maintenance of protected areas and conservation corridors of interest to tourism. Managers of

public protected areas often are the most effective players for conservation benefits from tourism development.

- Allocation of tax revenues for the protection of biodiversity-based tourism attractions, such as national parks and reserves.

In many biodiversity hotspot coun-tries, tourism destinations are under the influence of various governmen-tal agencies, whose mandates include and forestry. Smooth coordination among these departments and coherence between tourism policy and other government policies, including biodiversity conserva-tion, are not always the rule, and therefore different policies may undermine rather than support each other. Although tourism may be driven by the private sector, government policy instruments, such as requirements for environmental impact assessments (EIAs) and management plans, can be extraordinarily effective in ensuring that development takes place in an appropriate manner.

In Cancun, Mexico, for example, the Mexican government was criticized for "overlooking" zoning regulations and other development control mechanisms. Recently, however, with considerable pressure from outside sources, the government halted the construction of a resort complex on land owned by the developer because of its proximity to a sea turtle nesting area.

At the destination level, local authorities are often responsible for implementing policies regarding tourism and biodiversity conservation. Local authorities are well placed to negotiate between the various interests of local and outside entrepreneurs, civil society, and national government agencies, and they hold essential regulatory and zoning mandates that allow for the enforcement of guidelines and standards.

On the other hand, the capacity of local authorities to manage this complex and frag-mented industry effectively and to ensure its positive contribution to local strategies for sustainable development (for example Local Agenda 21 processes) is dependent on. whether local policymaking is coherent with national policy instruments and agencies.

Finally, it is noteworthy that in many economies in transitional and developing countries, national and regional governments often play the role of tour operators and hotel managers, either to try to jump-start quality standards or to generate revenue (for example, METS is a government-run tour operator in Suriname, and there are similar state-run operators in China and Vietnam). In many destinations, therefore, effects on biodiversity are crucially related to public policymaking and strategies.

**Recommendations for Governments**

- Use a multi-stake holder participatory planning process to develop national and local tourism strategies, policies, and master plans that reflect concerns about biodiversity conservation and poverty reduction. Integrate these into broader sustainable development strategies and processes, including traderelated poli-cies, investment promotion, economic incentives for the use of environmentally sound technologies, land-use planning, and taxation.
- Support private sector voluntary initiatives in conservation and provide opportunities for the private tourism sector to contribute to sustainable tourism management initiatives (community awareness and training, protected areas, etc.) through direct donations, in-kind services, and the estab-lishment and maintenance of private reserves.
- Enforce existing laws and regulations to avoid inappropriate development of tourism in core conservation areas, and avoid perverse incentives for tourism development to be environmentally damaging (such as reduced land taxes for cleared land).
- Control the planning, siting, design, and construction of tourism facilities and infrastructures according to biodi-versity conservation principles and guidelines.
- Improve awareness and exchange of knowledge between those responsible for and affected by tourism and nature conservation at a national, subnational, and local level.
- Undertake carrying capacity and limits to acceptable change assessments for sensitive areas and implement visitor-management plans based on assessment results.
- Develop or adopt certification schemes, reflecting national and local priorities that include biodiversity criteria, and pro-vide appropriate incentives for their adoption.
- Earmark adequate sources of funding for the management of natural areas. These funds should cover activities including protection of vulnerable ecosystems, management of visitor numbers, and support for surrounding communities.
- Develop management strategies, pilot projects, and mechanisms for sharing revenue from tourist visitation with the management authorities of protected areas, while keeping main management expenses covered by appropriate budgetary allotments.
- Promote and develop educational programs to enhance

awareness about nature conser-vation and sustainable use of biodiversity.
- Conduct market-assessment studies to avoid tourism facility construction in sensitive ecosystems that proves to be unfeasible and unnecessary.

## PRIVATE SECTOR

The tourism industry is character-ized by a large number of small and medium-sized enterprises (SMEs). At the same time, a significant amount of control within the industry rests in the hands of a very few, increasingly vertically integrated, multinational corporations. In Europe, for example, five companies control over 60 percent of organized outbound travel from the region. Few of these big companies have any long-term investments in particular destinations—even large hotel chain properties are often franchises rather than freehold properties. As such, their influence on tourism in a particular location may be much greater than their long-term commitment to that destination.

If environmental conditions worsen beyond a certain degree, these players potentially have the option of moving elsewhere. Furthermore, only a limited number of tourism companies have integrated biodi-versity considerations into their day-to-day management practices, and many remain unaware of the potential (and actual) impacts of their activities.

However, some major travel companies recognize the importance of managing their businesses to minimize their negative impacts and to find ways to help promote conservation and sustainable development. These companies realise that by helping to maintain the cultural and biological integrity of the places they visit, they can both enhance the quality of the product they are selling and improve their business reputation.

A significant development in the last few years is the establish-ment of voluntary environmental initiatives by hotel chains, tour operators, and ground handlers, including green certification systems, conservation awards, and ecolabels. While some of these initiatives are supported by governments and NGOs, all voluntary performance standard-setting depends essentially on private sector commitment and consumer awareness.

Initiatives such as the World Legacy Awards by Conservation International and National Geographic Society, British Airways' Tourism for Tomorrow Awards, the Green Globe 21 certification system, Australia's National Ecotourism Accreditation Programme, Co-opertive Research Center's International Ecotourism Standard, and the International Hotels Environment Initiative's benchmarking tool all provide guidance and added incentives for corporate responsibility toward biodiversity conservation.

### Outbound tour Operators and Ground Handlers

Assembling the component parts of holidays and managing significant numbers of tourists, outbound tour operators play a significant role in making a destination successful. Their capacity to manage their supply chains, their ability to ensure steady flows of tourists, and their ability to influence consumer choices make them key players in tourism development and biodiversity conservation. It is often the tour operators working in a specific destination that influence what type of tourism will develop there, what products will be available to the tourist, and even, in some cases, how much tourists will pay for these products.

Ground handlers often have a strong commitment to protect their destination because they are permanently based there, but their primary business concern is responding to the demands of the outbound tour operators they service. Therefore, there is a vital role that international tour operators can play in encouraging their suppliers to adopt environmentally and socially responsible management practices. The work of the Tour Operators' Initiative (TOI) for Sustainable Tourism Development noted below is a good example of how tour operators can work together to support biodiversity conservation efforts in the destina-tions where they operate.

### Accommodation Providers

Hotels, resorts, and other accom-modation facilities are both the tourism industry's main job generators and the main resource users (water, energy, land) affecting biodiversity in hotspot destinations. They also require a significant amount of infrastructure, such as roads and facilities for water supply and treatment, which if improperly developed can cause significant harm to biodiversity. However, hoteliers are increasingly recognizing the importance of maintaining the ecological integrity of the areas in which they operate.

CI estimates that approximately 8 percent of major hotel chain properties are located within the biodiversity hotspots. Maps, however, illustrate the actual and potential growth of the hotel industry in hotspot countries. Larger resorts are significant to conservation because they control large land properties and contribute significantly to tourism related employment. Their decision making is influenced by their complex ownership and management structure, involving asset owners, holding companies, and franchise/management corporations. However, the majority of hotels in hotspot destinations are independently owned, medium-sized enterprises.

**The Tour Operators' Initiative (TOI) for Sustainable Tourism Development**

The TOI is a network of 25 tour operators that have committed to incorporating sustainability principles into their business opera-tions and working together to promote and disseminate practices compatible with sustainable development. TOI members are taking action in three key areas:

- Supply chain management—to develop a common approach and tools for assessing suppliers.
- Cooperation with destinations—to exert a positive influence and speak with a collective voice on the actions of all partners, tour-ist boards, customers, suppliers, governments, and developers.
- Sustainability reporting—to develop and test reporting guidelines and performance indicators on sustainable development.

The TOI was developed by UNEP, UNESCO, and WTO, with technical and financial assistance provided by CELB. TOI is coordinated by a secretariat and hosted by UNEP, which ensures the implementa-tion of the programme of activities and continuous support to the members.

Small, medium, and large-scale hoteliers, with their assets tied up in a particular site, have a strong stake in the long-term sustainability of the surrounding environment. For this reason, the accommodation sector is often proactive in terms of social and environmental corporate account-ability and community outreach. However, the majority of environmental management programs for hotels located in biodiversity hotspots are focused on procedures for internal resource conservation and their related cost-savings, such as washing towels every other day, water saving, recycling, switching off lights, and so on.

Broader biodiversity considerations, such as maintaining natural habitats, avoiding land clearance, and setting aside property for species protection or as private nature reserves are yet to be integrated into such programs. In addition, the many different forms of accommodation ownership increases the difficulties of disseminating information and practical tools on how and why to incorporate sustainability practices and principles into their management. The International Hotels Environment Initiative has produced excellent materials for their member hoteliers, and others, which help integrate biodiversity considerations into hotel-management procedures.

### Cruise Ships

Since 1980, the cruise ship industry has had an annual growth rate of 8.4 percent and has grown nearly twice as fast as world international tourist arrivals over the past decade. Much of this growth is occurring in destinations that are located in the biodiversity hotspots. About 70 percent of cruise destinations are in the hotspots, including the Caribbean, the Mediterranean, Mexico, the Panama Canal Zone, and the South Pacific. V isits by cruise ships generate financial benefits to attractions, restaurants, retail shops, shore excursion operators, and other businesses at ports of call.

However, some cruise lines have had a past record of illegal waste discharge at sea. With predictions of further rapid growth over the next few decades, it will be increasingly important to understand and address the environmental impacts of cruising.

The cruise industry faces a number of key environmental challenges related to its activities and opera-tions in the world's oceans, particularly in and around priority conservation areas.

There is significant potential for wide-ranging negative environmental impacts from such hazards as mishandled waste and pollutants to poorly planned and implemented management processes. Several major cruise companies have done much to respond to the challenge of preserving the environment on which their business depends and are implementing leadership practices, testing and refining new technologies, and developing management programs to address environmental impacts.

The Caribbean, one of the most biodiversity-rich marine hotspots, accounts for 47 percent of the global yearly total of 54 million cruise ship bed days. The Mediterranean, another hotspot, is second as a cruise destination, with 12 percent of all cruise ship bed days. Construction of cruise ship ports and related infrastructure has a significant impact on coastal areas, and the building and maintenance costs are often borne by local governments with little means to design conservation-friendly facilities. The Centre for Environmental Leadership in Business at CI has launched a Cruise Ship Initiative to work with the cruise industry to reduce their environmental footprint and contribute to conservation in these key biodiversity hotspots.

**Air Travel**

Air carriers and related industry sectors affect biodiversity through their influence on airport siting and design and on destination-development decisionmaking (major tourism destinations are clearly dependent on the availability of an airport). Arguably, though, the biggest threat is their contribution to climate change: around 5 percent of global carbon emissions are attributed to air travel, according to the UN Intergovernmental Panel on Climate Change.

Climate change, in turn, through its impact on biodiversity—for example, recent coral reef bleaching episodes have been linked to changing weather patterns—indirectly affects biodiversity-based tourism. Some impacts are more direct, such as the potential loss of prime coastal sites and small islands associated with sea-level rise.

**Trade Associations**

Much of the private sector is organized into professional asso-ciations—at the global level, for example, the World Travel and Tourism Council (WTTC), the International Federation of Tour Operators (IFTO), the International Hotel and Restaurant Association (IHRA), and the International Council of Cruise Lines (ICCL), with scores of others at regional and national levels.

These associations have a significant ability to influence biodiversity impacts, and changes here are likely to result in a steep change across the industry.

Several associations, including those mentioned above, have already launched sustainability initiatives, some of which include limited support to biodiversity conservation. At the regional level, the Pacific Area Travel Association (PATA), the Caribbean Tourism Organization (CTO), and the Caribbean Hotel Association (CHA) have demonstrated a proactive attitude toward sustainability and conservation of natural resources.

***Recommendations for the Private Sector:***

- Incorporate biodiversity conser-vation practices and principles into the design, planning, development, and management of tourism products and services and into supply chain management.
- Commit to industryled, voluntary initiatives that include criteria for biodiversity conservation and socioeconomic benefits.
- Innovate processes and applications through new technologies and partnerships to minimize impacts on

sensitive ecosystems and to contribute effectively to the conservation of biodiversity.

- Make a commitment to educate staff and customers about the impacts of tourism on biodiversity and on local, traditional, and indigenous people.
- Cooperate with governmental and nongovernmental organiza-tions in charge of protected natural areas and the conserva-tion of biodiversity. Ensure that tourism operations are practiced according to the management plan and other regulations prevailing in those areas, so as to minimize negative impacts while enhancing the quality of the tourism experience and contributing financially to the conservation of biodiversity.
- Support destination-manage-ment efforts that seek to minimize the environmental footprint of the tourism indus-try and contribute to ongoing conservation initiatives.

**Table. Tourism-Related Donor Projects—Active or in Planning (2002)**

| Type of Project | Latin Amer. & Caribb-ean | Europe & Eurasia | Africa Near East | Asia & | Total |
|---|---|---|---|---|---|
| Tourism in Title | 24 | 1 | 5 | 1 | 31 |
| Accommodation | 3 | | 6 | 1 | 10 |
| Environmental Protection | 25 | 5 | 25 | 28 | 83 |
| Urban Development | 4 | 1 | 1 | 10 | 16 |
| Water Supply & Sanitation | 2 | 1 | | 1 | 4 |
| Agriculture | 2 | 1 | 2 | | 5 |
| Multisector | | 1 | 2 | | 3 |
| Social Protection | 1 | 4 | | 2 | 7 |
| Education | 2 | | | | 2 |
| Transportation | 2 | 1 | | | 3 |
| Microenterprises | 3 | | | | 3 |
| Health | 1 | | | | 1 |
| Science & Technology | 1 | | | | 1 |
| Utilities | | | | 1 | 1 |
| Private Sector | 1 | 2 | 3 | 1 | 7 |
| Industry | | | 1 | | 1 |
| TOTAL | 71 | 17 | 45 | 45 | 178 |

## DEVELOPMENT AGENCIES

Governments set the rules or frameworks under which tourism is developed. The private sector drives the development process itself. The missing link between the two is the financing of tourism planning and development. Financing can come from private sources, from governments, or from multilateral and bilateral donor agencies and development banks (sometimes directly, sometimes channeled through governments).

**The role of Development Agencies**

A survey of 55 development agen-cies conducted in 2002 for UNEP found that although private invest-ments are driven by economic incentives, donor interventions in tourism are motivated by longer-term development objectives, including alternative local income, natural resource management, community participation, employ-ment generation, and coastal zone management.

The George Washington University, CI, and UNEP have developed a database including details of over 320 tourism-related projects to determine the amount of donor funds that have been channeled into tourism develop-ment at a regional level and the types of projects being funded. Investment of those projects totaled over US$7 billion over 5 years. Table examines 178 projects, either active in 2002 or at various stages of implementation after ini-tial approval on this date (in pipe-line). It shows that of projects with significant tourism components, those related to environmental protection are most prevalent.

It is noteworthy that only 17 percent of the projects surveyed actually include tourism in their title,10 although all have important tour-ism components. This seems to indicate either that development agencies do not fully recognize the importance of tourism as a sus-tainable development tool or that they are concerned about criticism resulting from the poor environ-mental and social track record of tourism development projects in the 1970s and 1980s. This lack of definition makes it more difficult to study the scope and volume of tourism-related environmental conservation projects.

Two additional responses to the same survey bolster the conclusion that development agencies do not view themselves as important in setting the stage for sustainable tourism development:

- A selected group of 35 key experts in sustainable tourism were asked to say which types of organizations were most likely to fund projects dealing with sustainable tourism in the future. The experts' opinion that bilateral and multilateral aid agencies and government entities will most likely take the lead on this subject.

- However, the survey also asked the experts, as well as officers from development agencies, about the relative importance of key decision makers. The results show that although both experts and development agency officers thought that the private sector was the most important, the experts thought development agencies were much more important than the officers themselves did.

These results suggest that although others may recognize their impor-tance, development agencies may themselves be unaware of their role in influencing actors on the ground during the critical phase of siting, land-use planning, design, and choice of technologies and materi-als. The strength of this influence is evident in an example from Brazil, where experience is now shaping future projects of the Inter-American Development Bank (IDB).

**Table. Importance of Stakeholders as Drivers of Sustainable Tourism**

| | Private Sector | Govern-ment | Dev. Agency | Inve-stors | Local Pop. | NGOs | Tour-ists |
|---|---|---|---|---|---|---|---|
| Experts | 80 | 71 | 44 | 58 | 58 | 21 | 53 |
| Dev. Agency Officers | 100 | 65 | 27 | 72 | 83 | 28 | 53 |

## Recommendations for Development Agencies

- Redesign project portfolio on tourism as a carefully planned tool for biodiversity conserva-tion and poverty reduction, and incorporate monitoring, evalu-ation, and reporting procedures based on relevant indicators for biodiversity conservation results from tourism development.
- Support long-term public edu-cation and awareness-raising about the impacts of tourism on biological diversity; collect and disseminate lessons learned and best practices from existing project portfolio.
- Develop, adopt, adapt, or apply, as appropriate, conservation guidelines when preparing, approving, and funding tourism development projects having potential implications for bio-logical diversity.
- Invest in training and capacity-building to enable local people to benefit from tourism development.

## Local Residents in Tourism Destinations

The negative environmental, cultural, and social impacts of

unsustainable tourism development have affected local people most acutely. Traditional communi-ties and indigenous people can play a major role in conserving biodiversity, but this has been acknowledged only recently, and important issues relating to par-ticipation, land and resource use, and democracy still need to be addressed in the context of tourism development.

Local authorities have an essential role as moderators and facilitators of empowerment for local communities.

Experience with top-down approaches to protected area management has demonstrated that, if they are excluded, local people can undermine biodiversity conservation efforts (for example, the Maasai spearing of wildlife in Kenya's Amboseli National Park to protest removal of their grazing and watering rights within the park). Likewise, approaches to tourism development that do not take local people's priorities into account can be undermined by civil unrest and insecurity.

On the other hand, some of the successful examples of sustainable tourism development arresting or reversing biodiversity losses come from destinations where local authorities led feasible, multistakeholder governance systems (Calvia in Spain, Puerto Princesa in Philippines, Bonito in Brazil).

Local people often make up a large part of the workforce in the tourism industry, and labour organizations in the tourism sector have contributed to the debate on sustainable uses of biodiversity. The International Confederation of Free Trade Unions and the International Union of Food, Agricultural, Hotel, Restaurant, Catering, Tobacco, and Allied Workers' Association (IUF) address sustainable agriculture in rural areas and biotechnol-ogy.

Overall, the role of tourism employees in biodiversity conserva-tion is very important and should be considered in any global action plan.

Finally, local landowners play a crucial role in conservation, especially in buffer zones of core protected areas and in conservation corridors. The long-term survival of key ecosystems and species in a hotspot destination is often dependent on the land-use patterns around relatively pristine forests and coral reefs. Many biodiversity hotspot countries, such as Brazil, Costa Rica, and South Africa, have already established regulations, fiscal and economic incentives, and other policy tools to encour-age landowners to declare private reserves for direct or indirect use. Costa Rica's association of private reserve owners facilitates informa-tion exchange, promotes economies of scale, and lobbies government for additional support.

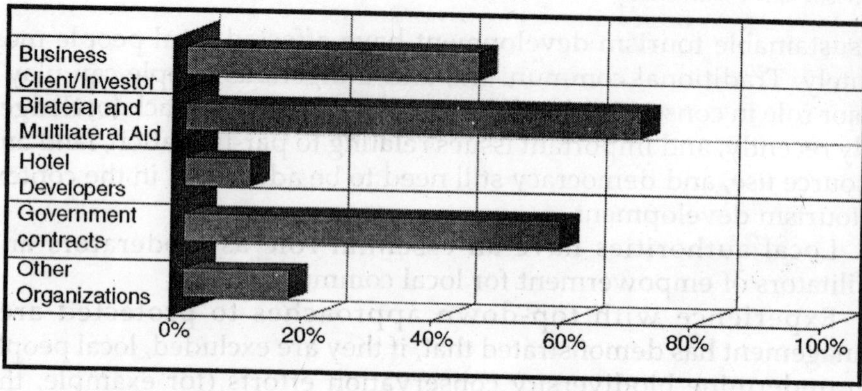


**Fig.** Projected Funding Sources for Sustainable Tourism Development.

## Recommendations for Local Residents in Tourism Destinations

- Establish representative governance systems that allow local people to be accountable and assume responsibilities in tourism and conservation partnership, and take action to fulfill them within the duration of the partnership.
- Identify, prioritize, and manage critical conservation area networks with direct and indirect tourism use at the local level, using tools such as impact assessment, market studies, zoning, and sustainable use plans.
- Be actively involved in and benefit from community capacity-building efforts relating to local sustainable tourism and ecotourism initiatives, in accordance with the principles of prior informed consent.
- Support local tourism contributions, such as traditional building techniques and materials; modes of transport; traditional foods, medicines, and handicrafts; and respect access to cultural sites.
- Collaborate with visitor-awareness campaigns on biodiversity conservation, educating tourists on the significance of natural resources for the local culture and the economy.

## OTHER MAJOR PLAYERS

### Consumers/tourists

Tourism businesses, like all other businesses, are dependent on consumers wanting to buy their products. A 1997 survey by the German association Studienkreis fuer Tourismus and Entwicklung found that 40 percent of German holidaymakers would spend an additional dollar to help save the environment in their destination—an estimated potential US$750 million. In 2000, Tearfund, a U.K.-based NGO, commissioned market research into consumer attitudes toward responsible tourism. The results showed that the U.K. tourist wants to relax on holiday, but not at the expense of local people or their environment. In practice, however, consumers have brought little pressure to bear on tourism companies.

Travel guides, magazines, and newspapers can make huge contributions to raising consumer awareness about critical issues facing the tourism industry and help to stimulate a demand for change. For example, National Geographic Traveller magazine routinely highlights issues of sustainable tourism and often profiles tourism businesses that are leading the way in implementing responsible travel practices. In 2000, the International Ecotourism Society (TIES) launched a travel media/tourist campaign, "Your Travel Choice Makes a Difference," which calls upon travel consumers to support tourism businesses that adhere to ecotourism principles and practices. At the same time, Audubon magazine developed a "Tread Lightly" code of conduct for travel in natural areas.

**Recommendations for Consumers/tourists**

- Respect local codes of conduct and visitor-management plans in sensitive areas.
- Ask tourism companies about their environmental and social standards and ensure they understand that your choice is determined by those standards.
- Actively support tourism busi-nesses and NGOs that are biodiversity friendly and seek to benefit local people.
- Recognize the cultural and natural diversity associated with many natural areas, particularly regarding local and indigenous communities.

**Nongovernmental Organizations**

Tourism has attracted considerable attention from international and local NGOs. Both conservation and development organizations have intervened in tourism with different, although often overlapping, objectives. At the international level, development organizations such as Tear fund and the International Institute for Environment and Development (IIED). Conservation organizations, including CI, the World Wildlife Fund (WWF), The Nature Conservancy (TNC), and the World Conservation Union (IUCN) are interested in tourism as a mechanism for biodiversity conservation, particularly using priority-setting criteria such as hotspots and eco-regions. These objectives are not mutually exclusive: Biodiversity conservation is part of sustainable development, and sustainable local livelihoods are critical to the success of biodiversity conservation. In other cases, NGOs such as Tourism Concern in the United Kingdom, Equations in India, and the Instituto de Hospitalidade in Brazil have been developed solely to focus on tourism and to promote a more responsible approach by the industry.

Local NGOs are at work throughout the developing world assisting communities to diversify their income through sustainable tourism and to protect the natural resource base.

They play an essen-tial role in pressuring governments and donors, raising awareness, mediating negotiations, building local capacity to deal with impacts of tourism, and implementing sustainable tourism projects on the ground. During the International Year of Ecotourism, UNEP identi-fied 35 NGOs in the South that had been active in the field of tourism and biodiversity for at least 3 years.

Some expressed concerns about tourism and its role in eco-nomic development—specifically about the expansion of nature and adventure travel into new (previously untouched) areas and the risk of damaging natural resources and the livelihood of indigenous communities as a result of poorly planned tourism growth. They say that for tourism to significantly contribute to sustainable development there must be proper man-agement and monitoring of such activities.

**Recommendations for Nongovernmental Organizations:**

- Facilitate and mediate innovative, conservation-friendly tourism partnerships.
- Play a role in the capacity building, technology transfer, and training of local communities to participate in and benefit from sustainable tourism development.
- Monitor impacts of tour-ism developments related to all stake-holders, and report independently and openly on results.
- Work with governments to integrate sustainable tourism as a biodiversity conservation strategy in national biodiversity agendas.
- Work with the private sec-tor to transform practices to more directly contribute to biodiversity conservation and benefit local and indigenous communities.
- Raise awareness among tourists as to the potential impacts—positive and negative—of tour-ism activities.
- Encourage informed decision- making among all stakeholders

regarding tourism development in or near local and indigenous communities.

- Fill research gaps on the dynamics of the relationship between tourism development and biodiversity conservation at both local and regional levels, and share best practices.

**Intergovernmental Organizations**

Through their influence on national governments, donors and lenders, and the secretariats of multilateral agreements related to tourism and environment, inter-governmental organizations such as UNEP, WTO, and UNCTAD play an important role in shaping tourism development by providing technical assistance and information, guidelines, facilitating nego-tiations, mediating agreements, and providing financial and logisti-cal resources.

Regional bodies such as the Organization of American States (OAS), the Caribbean Tourism Organization (CTO), and the Association of South East Asian Nations (ASEAN) have also developed guidelines, codes of ethics, and sets of principles.

**Inter-American Development Bank Lessons on Tourism Development With Conservation**

The Brazilian state of Bahia harbors one of the most threatened conservation hotspots, the Atlantic rain forest. The US$400-million PRODETUR I project, funded by the IDB from 1994 to 2001, improved and expanded eight international airports, built and improved over 800 kilometers of highways and access roads, provided water and sewage infrastructure, and attracted over US$4 billion in private tourism investment.

Its negative impacts on the environment, though, became clear to the bank officers: uncontrolled settlement of people looking for jobs, private building in environmentally sensitive areas, encroachment on rain forests and mangroves, and impacts on coastal reefs and other coastal ecosystems.

Intense pressure from local and international NGOs and community groups, supported by bank officials, ultimately overcame the initial resistance from investor groups and development-oriented government officers to allocate funds for conservation.

The result was the conservation of 22 historical heritage sites and the beginning of efforts to conserve over 70,000 hectares of coastal ecosystems and protected areas, including the creation of the new Serra do Conduru State Park. These lessons are being applied to new IDB projects in the region.

**Recommendations for Intergovernmental Organizations:**

- Assess and monitor biodiversity impacts of tourism development (and their social and economic determinants)—from supply through production to consumption; this assessment should assist in identifying effective intervention areas.
- Develop pilot projects (and research existing ones) to establish guidelines to support development and implementation of conservation-friendly tourism policies by governments and local authorities. These policies should address how to balance conflicting economic, social, and environmental priorities.
- Encourage transparent reporting of conservation issues by tourism corporations, support the development of certification and accreditation programs that con-sider conservation, and establish sector-specific sustainability reporting guidelines and performance indicators for biodiversity conservation.
- Work with key players and intermediaries to develop and implement customer awareness campaigns addressing conservation of biodiversity.
- Raise awareness of intermediaries on their key role in catalyzing change in the tourism supply chain, and build their capacity to improve conservation of biodiversity through networks and management tools (for example, screening indicators or environmentally sound technologies).

## Experts (Academics, Consulting Firms)

Caught between the development agencies and the private sector, experts and consulting firms (including the international accountancy and auditing companies) often serve as the implementing bodies for tourism development projects in developing countries—for example, in master planning and other national tourism development projects. In this sense, they often affect how and when tourism is implemented. Most strategies aimed at the private sector still overlook the potential role this sector can play in helping the industry become more biodiversity-friendly.

### *Recommendations for Experts*

- Incorporate biodiversity and socioeconomic considerations into tourism master plans and other development strategies.
- Ensure widespread participation in tourism planning

exercises, including biodiversity-focused stakeholders such as protected-area managers, NGOs, and traditional communities.

## TOURISM DEVELOPMENT, BIODIVERSITY CONSERVATION, AND LOCAL LIVELIHOODS—THE OVERLAP

Over the past 3 decades, major losses of virtually every kind of natural habitat have occurred, and the decline and extinction of species has emerged as a leading environmental issue. Many of the eco-systems in decline are the very basis for tourism development—coastal and marine areas, coral reefs, mountains, and rain forests—and support a wide range of tourism activities, including beach tourism, skiing, trekking, and wildlife view-ing.

The maps created in this study help to show the overlap between tourism development and areas of high biodiversity and threat—the "hotspots."

Biodiversity is essential for the continued development of the tourism industry, yet this study indicates an apparent lack of awareness of the links—positive and negative—between tourism development and biodiversity con-servation.

Tourism will require careful planning in the future to avoid having further negative impacts on biodiversity. Many of the factors associated with biodiversity loss—land conversion, climate change, pollution—are also linked to tour-ism development.

As the maps have shown, resources that are important for conserving biodiversity and supporting the livelihoods of local people, such as fresh water, are also affected by tourism development, and recognizing these links is important to managing tourism development into the future.

At the same time, an increasing number of examples have shown that tourism development guided by the principles associated with ecotourism—environmental sustainability, protection of nature, and supporting the well-being of local peoples—can have a positive impact on biodiversity conserva-tion.

By creating private reserves, providing justification for existing and new national parks and protected areas, and building a local conservation constituency among the people who live closest to important biodiversity areas, tour-ism can have and has had a positive impact on biodiversity conservation. Properly managed, tourism does have the potential to contrib-ute to biodiversity conservation and to support poverty reduction.

However, the proximity of tourism development to high-biodi-versity areas means that if it is not carefully managed, tourism could exacerbate the already rapid decline of biodiversity.

It is increasingly clear that, where tourism development is not guided by principles that promote conservation of nature and contribute to the well-being of local peoples, both human welfare and biodiversity can seriously suffer as a consequence.

38 hotspot countries that already experience more than 1 million annual tourism arrivals and lists 22 hotspot countries where tourist arrivals have more than doubled in the last decade.

In these, as well as in other biodiversity rich countries, incorporating biodiver-sity considerations into tourism development policies and strategies should be a priority. Although biodiversity was one of the five major issues under discussion at the 2002 WSSD in Johannesburg, tourism was not a specific agenda item.

Nevertheless, reference to tourism is made in the WSSD Plan of Action, including the need to continue ongoing work under the Convention on Biological Diversity on tourism as a sustainable use of biodiversity.

The emphasis is on the need to increase the benefits from tourism for host communities, while enhancing the protection of natural heritage and ecologically sensitive areas.

Proper management of tourism to ensure both social and environmental benefits requires collaboration among the "sustainable development triad" of public agencies, the private sector, and local communities.

Yet this study shows that the tourism industry and public authorities still appear to be largely unaware of the mutually supportive nature of the relationship between biodiversity conservation and tourism development and of the potential to use nature tourism as a way to contribute to biodiversity conservation efforts and poverty reduction.

The recommendations developed through this study for the key players (government, pri-vate sector, development agencies, local authorities, residents, and tourists themselves) can apply to both tourism destination countries and to the main tourist originating countries.

Acting on these recommendations will help to promote more responsible tourism activities and development that can positively affect both biodiversity and local economies.

During the past 10 years, and with the advent of ecotourism principles and sustainable tourism guidelines, significant progress has been made among the key players, both in terms of strategies for tourism development and in terms of collaboration to form more effective partnerships that consider environmental and human factors in designing tourism development plans and polices.

But as tourism continues to grow and expand into new natural areas, more concrete action needs to be put into place to safeguard the Earth's biodiversity, while promoting tangible economic benefits to local communities. It is our hope that the information contained in this study, by providing evidence of the overlaps and links between tourism, biodiversity, and local populations and the future implications of those overlaps, can contribute significantly to that effort.

# Chapter 5

# Biofuel Production and the Global Biofuel Industry

With the development of motorized vehicles in the beginning of the 20th century, visionaries such as Henry Ford and Rudolf Diesel advocated for producing liquid transportation fuels from biological materials. But because public policy and extensive research and development supported cheap fossil fuels, these bio-based industries never came to fruition. The oil shocks of the 1970s, however, once again brought both alcohol (ethanol) and combustible oil seed extracts (biodiesel) to the public's attention.

Brazil capitalized on its large sugarcane industry in the 1980s and began producing sugar based alcohol for automobile use. At the same time in the United States, corn-based alcohol was blended with gasoline to produce "gasohol" (10 percent ethanol) for cars. Political and industrial support for alternative fuels increased modestly until recently. Governments around the world have now created financial incentives for biofuel due to factors including industry lobbying, increased fossil fuel prices, climate change concerns and the desire for energy independence.

## BIOFUEL FEEDSTOCKS

Biofuel currently falls into two categories: ethanol, which is compatible with gasoline engines, and biodiesel, which is compatible with petroleum-based diesel engines. Many plant species can provide suitable biofuel yields. Ethanol is currently made from two basic feedstocks: starch-based feedstocks, such as corn, grain, wheat, barley and grain sorghum; and sugar-based feedstocks, such as sugarcane, sugar beets, fruits, citrus molasses and cane (sweet) sorghum.

Biodiesel has more diverse feedstocks than ethanol. It is derived from oilseeds, primarily rapeseed (canola), sunflower, soy and palm oil, as well as animal fat and vegetable waste products. While these feedstocks currently dominate biofuel production, feedstocks will likely expand into more environmentally beneficial cropping systems in the near future. For example, pilot plants are being developed for cellulosic ethanol production, which uses cellulose rather than plant starches.

Potential feedstocks for cellulosic ethanol include dedicated energy crops such as grasses and trees, as well as residues from agriculture (wheat straw, corn stover, sugarcane begasse), industry (paper pulp, sawdust) and forestry. Scientists are also researching other liquid biofuel and feedstock options.

## NEW FEEDSTOCKS

A number of new biofuel feedstocks are currently being investigated or already cultivated, some of which are described below. Many of these crops are attractive because they provide more environmental benefits than traditional biofuel crops such as corn, soybean and palm oil. But any feedstock, if cropped in large-scale, monoculture operations and/or if it destroys additional biodiverse regions, creates environmental concerns.

### Mixed Prairie Grasses

Cellulosic feedstocks are the most abundant source of biomass worldwide and are expected to become economically viable in the future, although opinions on when range from a couple of years to well over a decade. In contrast to cultivating cellulosic materials in plantations or unsustainably harvesting them from forests, a diverse mix of perennial crops could offer tremendous biodiversity benefits, higher biomass yields and a significantly improved energy balance.

David Tilman and his colleagues at the University of Minnesota conducted research over the course of a decade with mixed prairie grasses. They compared the energy yields of 172 different patches of grass species grown in a mix of between one and 19 plant species. The research revealed that on formerly highly degraded agricultural land, the mixture of 18 species yielded 238 percent more energy than monocultural systems.

The highly diverse plots are more productive because of a more efficient utilization of water and nutrients, better pest control and better resilience to weather perturbations. They also provide the additional benefit of reducing atmospheric greenhouse gas emissions by sequestering carbon in the soil.

### Jatropha

Jatropha has received considerable attention recently as an oil-rich feedstock for biodiesel. Jatropha curcas, a hardy deciduous perennial plant species, is able to survive extreme droughts and can be grown in the poorest, rockiest soils with little maintenance. The oil of the jatropha tree is easily processed and only needs to be modified slightly to produce biodiesel.

It also provides an opportunity to produce an economically viable crop in poor growing conditions, and possibly even mitigate desertification. Jatropha also presents some challenges. The crop consumes considerable water, and is invasive outside of Mexico. The toxicity of jatropha reduces its proliferation, but it could still replace native vegetation and reduce biodiversity. These concerns prompted a ban on jatropha cultivation in Western Australia in 2006.

Investments in jatropha plantations are increasing rapidly in Africa, India, Indonesia and China. India's goal of 20 percent biofuel by 2011 would require 13 million hectares (around 32 million acres) of jatropha plantations.15 Indonesia is planning to plant 1.5 million hectares of jatropha by 2010, an area the size of its current palm oil plantations. In March 2007, the Chinese forestry administration publicized its goal to develop 13 million hectares of trees high in oil-content, including jatropha. The energy industry is also interested in jatropha: BP is funding a $9.4 million project in India to investigate its utilization. And Archer Daniels Midland, Bayer CropSciences and Daimler recently announced a joint effort to develop jatropha as biodiesel.

### Sweet Sorghum

Sweet sorghum has significant potential as an ethanol feedstock, especially in countries such as India, China and the Philippines. Sweet sorghum does not require the conversion of starch into sugar before the ethanol is produced. Additionally, sweet sorghum can produce both food and fuel, as ethanol can be produced from the stalk juices and humans can consume the grain. The production of ethanol from sweet sorghum is not yet economically feasible.

A disadvantage of sweet sorghum is that its simple sugars cannot be stored as long as starches such as corn, and therefore the ethanol needs to be produced shortly after harvest. In India, ICRISAT (International Crops Research Institute for the Semi-Arid Tropics) is developing hybrids to improve the performance of sweet sorghum, which could make it competitive with other biofuel crops.

### Algae

The high lipid content of many algae species has spurred interest in commercializing algae production as a biodiesel feedstock. Most algae have an oil content of 15-40 percent and provide a high yield of biomass in a small footprint. From 1978 to 1996, the U.S. Department of Energy's Office of Fuels Development funded a programme to develop renewable transportation fuels from algae, the main focus of which was the production of biodiesel from high-lipid-content algae grown in ponds, utilizing waste $CO_2$ from coal-fired power plants.

Over the course of this programme, tremendous advances were made in the science of manipulating the metabolism of algae and the engineering of microalgae production systems. Many challenges remain and more research needs to be done before it will be known whether algae can be used on a large scale for biofuel production. Algal farms would be very input-intensive and require considerable nitrogen and phosphorus to maintain growth. Additionally, algae production is management-intensive and requires light throughout the water profile, not just on the surface, to grow productively.

### Other

Cassava (also known as yucca or manioc), a thick-rooted shrub that is an important food source for millions of people in Africa and South America, is becoming a popular ethanol feedstock, especially in China, where 13.3 million tons of cassava are already cultivated to produce ethanol. Small grains such as barley and wheat are also excellent feedstocks for ethanol production. A challenge with the use of cassava or small grains is the potential for competition with food production.

## BIOFUEL PRODUCTION TRENDS

The global biofuel industry is expanding rapidly. Ethanol continues to dominate in sheer volume.

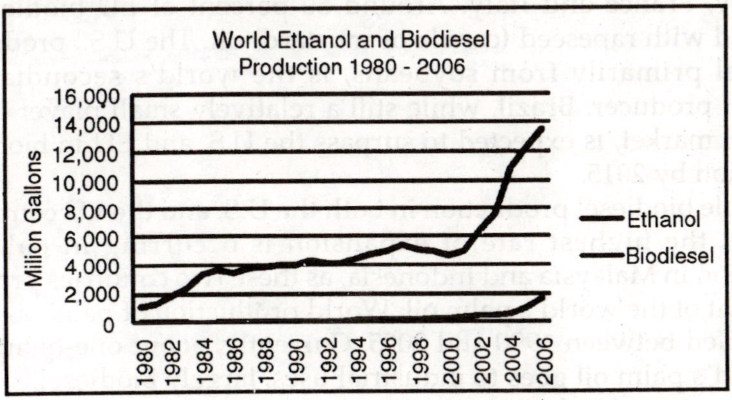


**Fig.** World Ethanol and Biodiesel Production

## ETHANOL

Currently, the five top ethanol-producing countries are Brazil, the United States, China, India and France. Brazil and the U.S. heavily dominate the market, using sugarcane and corn, respectively, to produce three-quarters of the world ethanol supply. China, the world's third-largest ethanol producer, has embarked on a National Fuel Ethanol Programme and, as of 2007, 10 Chinese provinces already have a mandatory 10 percent ethanol blend.

India, which produces ethanol from cane molasses, has mandated a five percent ethanol blend, although it has thus far been unable to reach that goal. The majority of ethanol is consumed in the countries where it is produced.

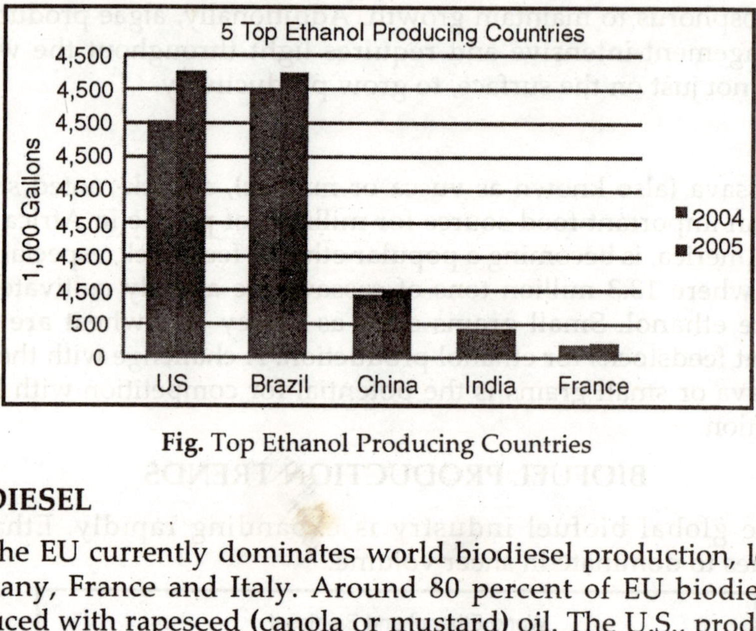


**Fig.** Top Ethanol Producing Countries

## BIODIESEL

The EU currently dominates world biodiesel production, led by Germany, France and Italy. Around 80 percent of EU biodiesel is produced with rapeseed (canola or mustard) oil. The U.S., producing biodiesel primarily from soybeans, is the world's secondlargest biodiesel producer. Brazil, while still a relatively small player in the biodiesel market, is expected to surpass the U.S. and EU in biodiesel production by 2015.

While biodiesel production in both the U.S. and the EU continues to grow, the highest rate of expansion is occurring in palm oil production in Malaysia and Indonesia, as these two countries produce 85 percent of the world's palm oil. World production of palm oil more than tripled between 1990 and 2005. Currently, about one-quarter of the world's palm oil goes to industrial uses, largely biodiesel.

Malaysia and Indonesia are also the largest exporters of palm oil, exporting primarily to the EU. Although it produces biodiesel itself, the EU is limited in cropland for feedstock expansion and has set aggressive targets for biofuel use.

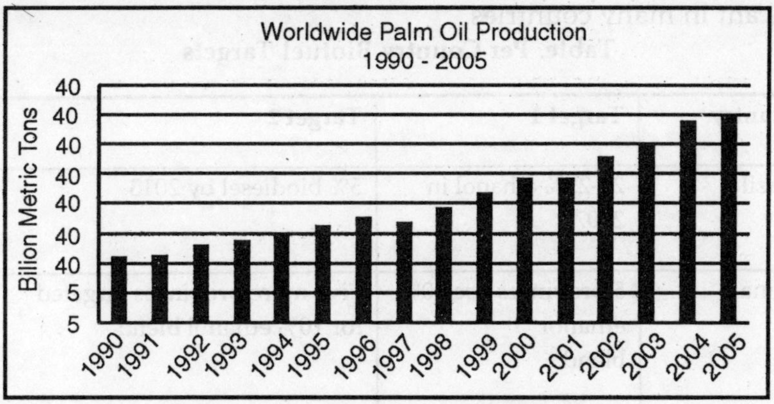


**Fig.** Worldwide Palm Oil Production

Other major palm oil-importing countries, though not necessarily for biodiesel, include China, India and Pakistan, with the U.S. holding only a minor share.

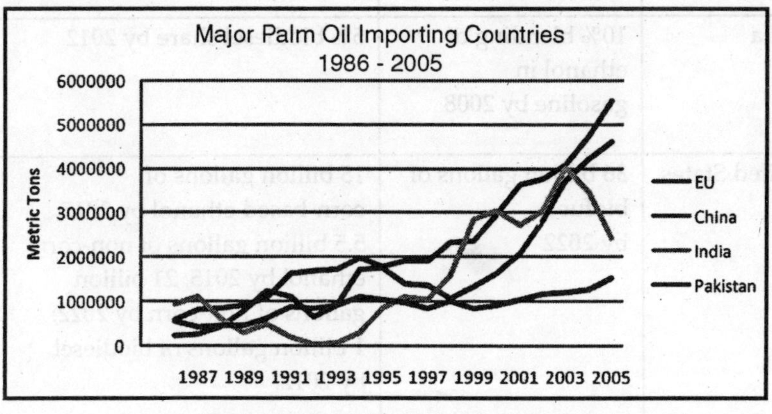


**Fig.** Major Palm Oil Importing

## CONTRIBUTING FACTORS: DOMESTIC POLICY INCENTIVES AND INTERNATIONAL TRADE RULES

The use of biofuel has been and continues to be motivated by a complex web of supports and incentives. Renewable fuels standards—government policies designed to carve out a share of the transportation fuel market for biofuel—have been adopted by several countries.

Other major policy drivers include government support for research and development of refineries and crops, subsidies directed to fuel blenders, low feedstock prices driven by agricultural overproduction, tax rebates and tariffs. Funding from international monetary organizations such as the World Bank has also been significant in many countries.

**Table. Per Country Biofuel Targets**

| Country | Target 1 | Target 2 |
|---|---|---|
| Brazil | 20-25% ethanol in 2007 | 5% biodiesel by 2013 |
| China | 5 provinces use 10% ethanol blend | Five more provinces targeted for 10% ethanol blend |
| EU 27 | 5.75% biofuel share of transportation fuel by 2010 | 10% biofuel share by 2020 |
| India | 10% blending of ethanol in gasoline by 2008 | 5% biodiesel share by 2012 |
| United States | 36 billion gallons of biofuels by 2022 | 15 billion gallons of corn-based ethanol by 2015; 5.5 billion gallons of non-corn ethanol by 2015; 21 billion gallons of non-corn by 2022; 1 billion gallons of biodiesel by 2012. |
| Indonesia | 10% biofuel by 2010 | |
| Malaysia | 5% biodiesel on public vehicles | |

## TRADE RULES AND BIOFUEL

International trade rules will influence how and where the global biofuel industry matures. While some countries seek to export biofuel feedstocks, other countries have created biofuel consumption targets that are not feasible using internally produced feedstocks, creating a substantial driver for biofuel trade. Trade in biofuel remains small. In 2004, about 3 billion litres of ethanol were traded internationally, compared to 920 billion litres of crude oil.

In 2005, about 10 percent of total biofuel consumption was traded internationally. This trade will expand, probably rapidly, as a result of the various targets created in several industrialized countries to create a minimum use for biofuels. The EU target of 10 percent biofuel by 2020 has drawn particular attention. Many governments in Europe are revising incentives for biofuel to protect the environment, and the EU is proposing to restrict imports of biofuel that has not been produced in a sustainable manner.

In the United States there is strong support for domestic fuel production, motivated by both agricultural interests (new markets for domestic producers) and a bid for energy-security (reducing imports of energy).

There is no separate framework of rules governing trade in biofuel. Following the designation made by the World Customs Union, the World Trade Organization (WTO) treats ethanol as an agricultural product, subject to the Agreement on Agriculture.

Biodiesel, however, is considered an industrial product, and is therefore subject to the Agreement on Subsidies and Countervailing Measures. Agriculture has been disciplined by multilateral trade rules only since the signing of the Uruguay Round in 1994.

Trade in agricultural goods remains full of exceptions to the rules that govern trade in other goods; agriculture rules allow relatively high levels of domestic support, some extremely high tariffs and, for now, the continued, if constrained, use of export subsidies by those WTO members that were using them when the Uruguay Round Agreements were signed.

Subsidies in the energy sector (such as U.S. tax breaks for the domestic petroleum industry) are likely WTO-illegal but have not been challenged, perhaps in part because a number of the world's major oil exporters are not WTO members (e.g., Russia) or have only recently acceded (e.g., Saudi Arabia). As an industrial product, biodiesel faces low tariffs in industrialized countries.

Biodiesel feedstocks, however, as agricultural commodities, are generally protected through agricultural support payments and tariffs. Oilseeds, many of which can be used to generate biodiesel, are an exception for the EU, which has an agreement in place to accept oilseeds duty-free. Given WTO norms and rules, it would be very difficult for members to introduce new, higher tariffs on biofuel, although other market access barriers exist or could arise related to standards.

## CURRENT BIOFUEL PRODUCTION

Current biofuel production based on corn, soybean, sugarcane and palm oil has direct and indirect impacts on biodiversity in all stages of its life-cycle: feedstock production, biofuel production and biofuel combustion. Perhaps the most effective way to explore the impacts of biofuel on biodiversity is to look at case studies where the industry is developing the fastest: Indonesia and Malaysia, Brazil and the United States.

## INDONESIA AND MALAYSIA

### Biodiversity

The tropical forests of Indonesia, which cover around 48 percent of the country's land area, are among the most diverse on the planet. Even though the country is very small, Indonesia hosts 10 percent of all plant species, 16 percent of all reptiles and amphibians, 12 percent of all mammal species and 17 percent of all bird species. Like Indonesia, Malaysia's tropical forests are also among the most diverse on the planet.

*Indonesian Wildlife in 2002*:

- *Breeding Birds*: 929 species, 114 threatened
- *Mammals*: 515 species, 147 threatened
- *Higher Plants*: 29,375 species, 384 threatened
- *Fish*: 4,080 species, 68 threatened
- *Amphibians*: 278 species
- *Reptiles*: 745 species, 28 threatened

*Malaysian Wildlife in 2002*:

- *Breeding Birds*: 245 species, 37 threatened
- *Mammals*: 300 species, 50 threatened
- *Higher Plants*: 15,500 species, 681 threatened
- *Fish*: 368 species, 20 threatened
- *Amphibians*: 198 species
- *Reptiles*: 379 species, 21 threatened

Nearly all of Malaysia and the largest part of Indonesia, including Sumatra and Borneo, are part of the Conservation International Sundaland Hotspot. The hotspot is unique in its diversity and hosts 15,000 endemic plant species and more than 160 endemic animal species. Altogether, Indonesia has around 2,500 known animal species (excluding around 4,000 species of fish) and Malaysia has around 1,500 known species of animals (including fish).

It is estimated that thousands of species in both countries are still undetected. Key species include the Sumatran tiger, the Sumatran rhinoceros, the gibbon, the clouded leopard and human's closest relative: the Bornean and the Sumatran orangutan.

## Biofuel Trends

Although the majority of palm oil is produced in Malaysia and Indonesia, the crop originated in central Africa where it remains an important commodity. It was introduced to Southeast Asia in the 19th century, but it was not until the 1960s that a vast monoculture of palm was established in Malaysia and only in the 1980s in Indonesia. Since the 1960s, Indonesia's palm oil plantations have grown 30-fold, and in 2003 they covered 12,000 square miles. Palm oil plantations in Indonesia increased from 150,000 acres to 1,000,000 acres between 1985 and 2006, an increase of around 660 percent. In Malaysia, plantations in 2003 covered 13,500 square miles, consuming 11 percent of Malaysia's total area and 62 percent of the country's cultivated agricultural land. Between 2000 and 2005, palm oil exports from Indonesia rose from 4.1 million tons to 10.4 million tons annually, while in Malaysia exports rose from 8.1 million tons to 13.2 million tons annually.

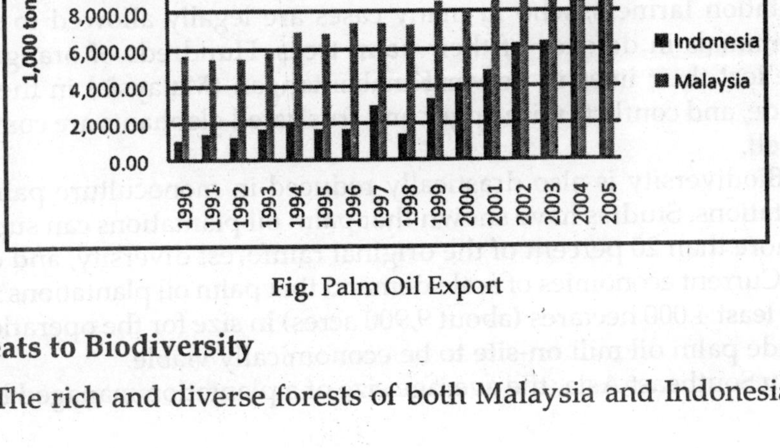


Fig. Palm Oil Export

## Threats to Biodiversity

The rich and diverse forests of both Malaysia and Indonesia are threatened by the clearing of land for commercial pulp, rubber and palm oil. According to a paper issued by Friends of the Earth, industry data indicate that by 2002, around 48 percent of all palm oil plantations in Indonesia and Malaysia involved forest destruction.

In Malaysia, 87 percent of forest destruction between 1985 and 2000 was connected to the increase of palm tree plantations. Between 1990 and 2000, Indonesia lost an average of 0.5 million acres of rainforest annually, with an annual 1.6 percent loss in forest cover. Between 2000 and 2005, the average annual loss increased to around two percent of forests, resulting in an average annual loss of around seven million acres.

It is estimated that if logging continues at the current rate, no lowland tropical forest will exist in Indonesia's Kalimantan province by 2010. Malaysia lost around 19,500 acres of forests annually between 1990 and 2000, an average deforestation rate of 0.35 percent per year. The annual deforestation rate jumped to 0.65 percent between 2000 and 2005, an increase of more than 85 percent.

In just 15 years, Malaysia lost 6.6 percent of its forest cover, or 375,000 hectares (around 930,000 acres). The islands of Borneo (split by Indonesia and Malaysia) and Sumatra (Indonesia) have already lost 50 percent and 70 percent of their original forest cover, respectively, and most of the loss has been attributed to logging companies and, more recently, palm oil companies.

Deforestation, habitat fragmentation and development threaten the survival of many species in the two countries. Many of the threatened species are clustered together in small patches of remaining forest stands. It is estimated that nearly 50 percent of Indonesia's forests are fragmented. Fragmentation not only destroys corridors but also exposes wildlife to illegal hunting and poaching.

In addition, animals in search for food often collide with poor plantation farmers, who in many cases are legally allowed to shoot the animals in defence of their palm trees. Hundreds of orangutans have lost their lives in Lower Kinabantangan (Malaysia) in the past decade, and conflicts with angry and frustrated elephants are common as well.

Biodiversity is also drastically reduced in monoculture palm oil plantations. Studies have shown that palm oil plantations can support no more than 20 percent of the original rainforest diversity, and often less. Current economies of scale demand that palm oil plantations must be at least 4,000 hectares (about 9,900 acres) in size for the operation of a crude palm oil mill on-site to be economically viable.

In Southeast Asia, the average size of a plantation managed by an individual plantation company is 10,000-25,000 hectares (25,000-60,000 acres), and many of these companies belong to even larger agribusiness holdings themselves. Slash and burn practices, deforestation and drying and burning of peatland also release large amounts of carbon dioxide and contribute to Indonesia's status as the third largest $CO_2$ emitter worldwide. Not only does this enhance global warming, it destroys precious ecosystems.

Leaching agrochemicals and sediments are also causing severe water pollution, leading to fish and coral kills. Paraquat dichloride, a toxic herbicide banned in several countries, is commonly used on palm oil plantations. Water consumption is also a concern, as the drying of peat lands and deforestation have sharply reduced the water- retaining capacity of the soil, thereby increasing runoff and disturbing hydrologic cycles in Indonesia and Malaysia.

### *Policy Supports*

Corruption and cronyism have certainly played a role and have often been responsible for unenforced logging bans, ignored environmental standards, lost forest fire battles and other problems. For instance, in its five-year plan, Indonesia allocates forest land, prioritizing degraded lands. Timber companies often purposely leave forests in poor condition so that conversion to palm plantations can be justified.

In addition, the government readily provides permits for forestland, even though large tracts of former agricultural land are set aside for that purpose. Forested land is more attractive than agricultural land because companies can first log the trees and then sell the timber.

### *What's Being Done*

Several initiatives are trying to address the negative impacts of palm oil production in Indonesia and Malaysia. To deal with unsustainable practices, the Roundtable on Sustainable Palm Oil (RSPO) was established in 2003 in Kuala Lumpur, Malaysia. Companies and environmental organizations, such as World Wildlife Fund, develop voluntary standards and guidelines for plantation managers. RSPO offers trainings for plantation managers and monitors the performance.

The RSPO has received increasing industry support from companies like Unilever, Johnson & Johnson, Nestle SA, H.J. Heinz Co. and Cargill. While the RSPO hopes to independently certify palm oil refiners in 2008, it has yet to certify any operation in Malaysia or Indonesia. Operations will have to prove they have not harmed the environment, and plantations on forested areas destroyed after November 2005 will not be eligible for certification.

The RSPO has been criticized by many NGOs for being dominated by industry and for not having sufficient representation from indigenous peoples, farmers, small land holders and workers, among other issues.

Furthermore, questions remain about the RSPO's willingness to address larger sustainability issues around genetically modified organisms (GMOs), mono-cropping and the effectiveness of the voluntary (as opposed to regulatory) nature of the standards.

The ecologically destructive practices associated with palm oil production have led the European parliamentary committee to recommend a total ban on palm oil imports until sustainable production practices are utilized. But at the same time, the EU has set a goal of 10 percent biofuel blends in all new vehicles by 2020.

If palm oil imports are banned, this would create a huge market for other biofuel feedstocks. Whether an EU ban will have a strong influence on the export markets in Indonesia and Malaysia is questionable, because many other global markets, including China and India, exist for palm oil.

### Goldten Hope Plantations, Berhad, Malaysia

Golden Hope Plantations is often used as an example of a plantation using more sustainable methods. These include the following:

- *No deforestation*: Establishment of plantations only on agricultural land or formerly deforested land, not on new forest and peat land.
- *Zero burning techniques*: Instead of burning the vegetation, the company shreds the material and allows it to compost, which reduces greenhouse gas emissions, controls erosion and adds nutrients to the soil, thereby reducing fertilizer requirements.
- *Strongly restricted use of pesticides*: The farm frequently uses biological controls, such as beneficial insects and owls for rodent control.
- *Use of palm oil mill effluent*: The effluent is treated through anaerobic digestion, the methane captured and the residues used as fertilizers.
- *Water usage*: The water used in irrigation systems is carefully managed.
- *Biodiversity*: Natural forest lands are restored and sustainably harvested for food and medicinal plants.

## BRAZIL

### Biodiversity

Brazil has some of the most diverse bioregions in the world, including six unique ecoregions: Mata Alantica, Cerrado, Amazonia, Pantanal, Caatinga and Pampa. Species in Brazil:

- *Higher Plants*: 56,215 species, 381 threatened
- *Mammals*: 394 species, 81 threatened
- *Breeding Birds*: 686 species, 114 threatened
- *Reptiles*: 648 species, 22 threatened
- Amphibians: 681 species, 6 threatened
- Fish: 471 species, 17 threatened

**Table. Unique Species per 10,000 km$^2$ Brazil**

| | Number of species |
|---|---|
| Mammals | 0.42 |
| Breeding Birds | 0.8 |
| Reptiles | 0.78 |
| Amphibians | 0.79 |

The Atlantic Rain Forest (Mata Atlantica) (including the Pampa, which is largely located in Argentina and Uruguay) is among the top five biodiversity hotspots on Earth. It contains more than 260 species of mammals, 450 amphibians, 930 birds, and 20,000 plants and is home to numerous endangered species, including the marmoset and golden lion tamarin.

The region also contains 55 endemic threatened birds, 21 mammals and 14 endemic threatened amphibians. The floral biodiversity is actually greater than that of the Amazon. The Cerrado biome, often referred to as a woodland-savannah, is the second largest of the country's biomes after the Amazon. It consists largely of woodland-savannah and dry forest ecosystems and has an estimated 160,000 species of plants, fungi and animals, 800 species of trees and large shrubs and a large amount of ground shrubs.

It is home to the marsh and pampas deer, the giant anteater, the puma and the giant armadillo. There are about 300 other mammals and 935 species of birds as well. The Amazon is the most widely known biome in all of South America. It represents more than half of the world's remaining rainforest and is the most species-rich rainforest in the world.

About 2.5 million insect species, more than 40,000 plant species, 3,000 fish, 1,294 birds, 427 mammals and 378 reptiles have been scientifically described in the Amazon. The Pantanal is the world's largest freshwater wetland, measuring almost ten times the size of the Everglades. There is an incredible diversity of flora and fauna that have adapted to the area's seasonal flooding, including 260 species of fish and 650 species of birds. Many North American birds also over-winter in the Pantanal.

The Caatinga of the Brazilian Northeast is one of the richest dry forests in the world. It has at least 1,200 plants, nearly 600 arboreal and shrubs, and many fish and amphibians. Two macaws, the indigo and the little blue, are among the 10 most endangered birds in the world.

## Biofuel Trends

Currently, more than 300 sugar ethanol mills operate in Brazil with nearly 90 new mills planned by 2014. In 2006, Brazil produced nearly 4.5 billion gallons of ethanol. Ethanol plants are currently concentrated in the state of Sao Paulo, in the Atlantic Forest biome. About 14 million acres are currently planted to sugarcane in Brazil, half of which are used for ethanol production and half for sugar production.

As of August 2006, Brazil also had five biodiesel plants in operation, with a total production capacity of nearly 13 million gallons per year (MMgy) and another five plants representing an additional 16 MMgy under development and/or construction. There are plans for another 24 plants with 264 million gallons of capacity in the future.

Currently, Brazil uses primarily soybean in its biodiesel production, along with some palm oil and castor oil, but is researching additional feedstocks. One biofuel plant in Mato Grosso, in the heart of Brazil's centre-west soybean region, integrates both ethanol and biodiesel production, with resulting energy and cost savings. Brazil's ethanol production currently accounts for more than 40 percent of the fuel needs of its gasoline-powered motor vehicles.

Considering the diesel mix in Brazil's fleet, this works out to about 12 percent of all its transportation fuels. As the global demand for biofuel grows, Brazil plans to further position itself as an exporter of biofuels.

## Threats to Biodiversity

As in Malaysia and Indonesia, deforestation—and its impact on biodiversity—has been one of the primary results of the biofuel boom in Brazil. The country's Atlantic Rainforest (Mata Atlantica) once covered more than a half million square miles, but today only about seven percent remains, and much of that is severely fragmented.

The Mata Atlantica is also one of the most intensely farmed areas in Brazil, home to more than 60 percent of the country's sugarcane and 85 percent of its ethanol production. The Cerrado has been extensively developed recently for soybean production as well as corn, rice, cotton, coffee and ranching. It is estimated that nearly 60 percent of the Cerrado's original vegetation has now been completely destroyed, and Conservation International predicts that the Cerrado could disappear by 2030.

While the demand for beef and soy protein has fueled the clearing of the Cerrado in the past, the demand for both sugarcane and soybean for biofuel likely will accelerate the loss of biodiversity in this region. Since 1990, the area planted to soybean in Amazonian states has expanded at the rate of 14 percent per year. Soybean, used for animal feed and increasingly for biodiesel, has become a major driver of deforestation. About 20 percent of the Amazon rainforest has been cleared in the past 40 years.

As Brazil's ethanol demand increases, sugarcane plantings are expected to expand. Western Bahia, which historically has had soybean, cotton, coffee and corn, now plans to expand into sugarcane. Between 20,000 and 30,000 hectares of sugarcane could be planted in the region by 2010. The expansion of sugarcane will likely displace other crops, including soybean, pushing soybean production to the north and into the Cerrado and the Amazon.

It is also possible that growing biofuel demand will push sugarcane into the Amazon itself unless the country devises sufficient safeguards. Water pollution from farming and sugarcane refineries has affected Brazil's rivers, particularly in the Pantanal. As soybean cultivation increased in the neighboring Cerrado, water draining from these fields has further decreased water quality. The Pantanal earned the dubious distinction of being named the "Threatened Lake of the Year 2007" by the Global Nature Fund.

If petroleum prices remain high, and if the U.S. removes its ethanol tariff of 54 cents per gallon, then enormous growth in Brazilian ethanol exports can be expected. This will put large pressures on the Amazon, even if it only intensifies the use of land that has already been cleared. The recent meeting of U.S. President Bush and Brazilian President Lula, during which cooperation in the biofuel sector between the U.S. and Brazil was discussed, highlighted the potentially dramatic new demand for Amazonian land.

**Policy Support**

Brazil emerged as an international leader in biofuel production through a combination of public policy, strong research programs, its excellent climate, and the low cost of land and labour in Brazil. After the global oil crisis in 1973, the Brazilian government decided to develop the ethanol industry. In 1975, a programme known as "Proalcool" was initiated to accelerate ethanol research and development.

The programme also provided subsidies for sales of ethanol as well as flex fuel vehicles. This programme faltered when global oil prices dropped, but demand rebounded with higher petroleum prices. Most recently, the government has passed legislation that mandates a two percent blend of biodiesel in all commercial sales of petroleum diesel by 2008, rising to five percent by 2013.

### What's Being Done

An attempt to help meet the European Union's renewable fuel mandate of 10 percent by 2020 is driving some of the biofuel expansion in the country. But the damage to biodiversity from growing more biofuel feedstock has drawn criticism from the EU, with proposals to restrict import of biofuel that has damaged the environment.

Additionally, EU officials have expressed concern that ethanol makers were breaking local environmental and labour laws, in particular the use of slave labour in sugarcane harvest and the destruction of tropical forests. In response, the Brazilian government is taking steps to require ethanol companies to obtain a new environmental and industrial standards certificate before ethanol can be exported in 2008.

## UNITED STATES

### Biodiversity

The U.S. "corn belt," where biofuel production is currently concentrated, has a number of different bioregions.

The Upper Midwest Forest-Savanna Transition, which is between the eastern forests and the northern edge of the Great Plains, contains a mosaic of forests, savannas and largely oak, maple and basswood woodlands. The oak savannah is one of the world's most endangered ecosystems.

Riparian areas are important to migratory birds, yet less than five percent of the ecoregion is intact, and the remaining habitat is highly fragmented. The Northern Tall Grasslands region is the northern-most extension of the tall grass prairie, extending from Lake Manitoba south along the Red River Valley to central Minnesota. The dominant grasses are big bluestem, switchgrass and Indian grass, which are often mixed with quaking aspen and oak groves, along with rough fescue grasslands.

The long growing season and abundant rainfall in the Central Tall Grasslands, which covers southern Minnesota, most of Iowa and small areas of eastern South Dakota, Nebraska and Kansas, has given it one of the most productive ecoregions on earth. Prior to agricultural development, bison and likely elk dominated the large animals, but the wolf also was important to the ecosystem.

There is a high level of fragmentation, and virtually no significant blocks of intact habitat remain. The Central Forest-Grasslands Transition extends through northern Illinois, Missouri, eastern Kansas, Oklahoma and Texas and has a mixture of savanna, prairie and woodlands. It was one of the richer ecoregions in North America with a large number of species of reptiles, birds, butterflies and tree species, dominated by oak and hickory.

As with the other central U.S. ecoregions, almost all of the area is now cultivated for corn and soybeans. Less than one percent of the native habitat is considered to be intact, and the degree of fragmentation is extremely high. The western-most part of the U.S. grain producing regions, the Central and Southern Mixed Grasslands, runs north-south from central Kansas through western Oklahoma and north-central Texas. It is a mixture of tall and short grass prairies and has a high floral complexity.

Today only about five percent of the remaining habitat is intact, and 90 percent of the area has been converted to cropland and pasture. High wheat prices led to conversion of much of the area in the 1990s, accelerated by use of centre pivot irrigation.

## Biofuel Trends

The corn-based ethanol industry in the U.S. is growing rapidly. U.S. ethanol production doubled between 2001 and 2005, and it will likely double again within the next few years. According to the Renewable Fuels Association, as of December 18, 2007, the U.S. had 135 active ethanol plants capable of producing more than seven billion gallons of ethanol per year.

An additional 65 plants under construction and nine plant expansions will add more than six billion more gallons of capacity, bringing total capacity to more than 13 billion gallons annually. Construction is expected to continue following the passage of a stronger renewable fuel standard by Congress in December 2007, which requires 36 billion gallons of renewable fuel by 2022.

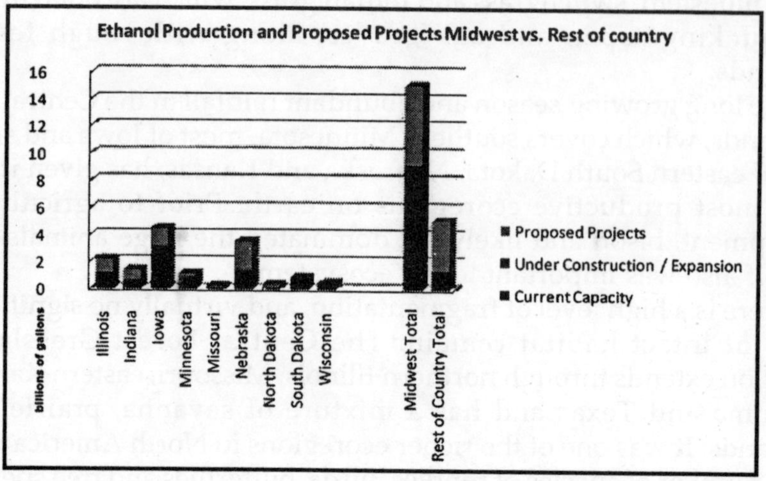


**Fig.** U.S. Ethanol Production and Proposed Projects

The amount of U.S. corn going into ethanol has also expanded rapidly, from about five percent per year from 1990 to the early 2000s to almost 20 percent in 2006. U.S. farmers planted more than 90 million acres of corn in 2007, up from nearly 79 million acres in 2006 and the most since the 1940s. Soybean biodiesel production is also significant. In 2007, it reached 250 million gallons, requiring 5.6 percent of U.S. soy production.

If all proposed soybean biodiesel plants came online, about 1.7 billion gallons could be produced, or about 42 percent of U.S. soybean production. While biofuel production has thus far been concentrated in the traditional Midwest "Corn Belt," the agricultural region is expanding as increasing demand for corn and improved genetics and management have adapted corn to more extreme climates and poorer soils. Corn has begun to displace wheat and sugar beets in the Dakotas and cotton in the southern part of the U.S.

Corn production has also intensified in the Midwest, in some cases replacing soybeans in the traditional corn-soybean rotation. Ethanol plants are being constructed around the country, including in states that produce little or no corn. Although the past few years have been extremely profitable for ethanol producers and investors, a number of uncertainties may slow the market, including fluctuations in petroleum prices, higher corn prices and pest pressures that may increase the costs of monocultural corn and soybean production.

## BIODIVERSITY IMPACTS

Overall, the native U.S. ecosystems have been so highly altered that native biodiversity has been largely eliminated, except for fragmented remnants and protected areas. While many of these areas were significantly altered before biofuel production began, some of this alteration is due to the production of the same crops that are now being used for biofuel—corn and soybeans. It is critical that these few protected areas be maintained.

The contiguous United States uses about half of its land for production of food, feed and fibre. While much is grasslands and forests, 17 percent is devoted to annual commodity grain crops, including corn, soybean and wheat. These crops dominate in part because of the development of efficient production systems, including machinery, hybrid seed, fertilizers and pesticides, as well as irrigation and drainage. When humans manage an ecosystem for the purpose of food, fibre and fuel production, they often limit the biodiversity of that land area.

However, the degree to which they limit biodiversity depends upon the type of agriculture that is practiced. For example, the state of Iowa has 32,000 square miles of corn and soybeans, representing 92 percent of all its farm acreage. By contrast, researchers have found more than 50 species per garden in home gardens in Mexico, and a combined 404 species across the home gardens in just one Mexican community.

Expansion of grains and oilseeds for biofuel and other uses may well continue to lower biodiversity and, importantly, limit the fledgling efforts at restoring habitat fragments. Examples include the possible loss of conservation reserve land. Nebraska has already seen a decline in conservation reserve land by about 6.5 percent between September and November 2007, from 1.34 million acres to about 1.25 million. Wetlands are one of the most important ecosystems in the U.S.

Wetlands have been disappearing for decades because of drainage for agricultural and development purposes, and they continue to lose their ability to provide ecosystem services, including biodiversity. The Prairie Pothole Region, which extends from northwest Iowa through western Minnesota, the Dakotas, Montana and into western Canada, has seen a major loss in wetlands. These shallow wetlands were formed from depressions left after the last glaciations and are critical to the migration of many wild birds. Many have been protected by federal land preservation agreements, and the U.S.

Fish and Wildlife Service (FWS) has been buying these wetlands for preservation as rapidly as funds permit. However, land prices have increased, stretching the FWS funds thin, and farmers are not renewing government contracts because of the high price of corn and soybeans. The economic pressure coupled with loss of refuges due to global warming indicates that this critical ecosystem may suffer even further loss. High-nutrient (fertilizer) and pesticide inputs, as well as accelerated soil erosion, also impact water supplies and lower soil quality.

The nutrient-driven hypoxic zone (dead zone) in the Gulf of Mexico reached a record 7,900 square miles in 2007, one of the largest on record. In addition, consecutive plantings of corn require more nitrogen fertilizer than when corn is alternated with soybeans. If corn production continues to expand in the U.S., so too will the impacts on the environment.

## Policy Supports

In the U.S., biofuel policies were initiated in the 1970s and have been reinforced with subsidies and renewable fuel standards in the past few years. U.S. farm policy also plays a significant role. While farm policies initiated in the 1930s were devised to maintain a fair market price for commodity farmers, these policies have largely been abandoned in favour of policies that drive down prices, often well below the cost of production.

The very low price of corn over the past 30 years has been an underlying driver behind the crop's rapid expansion. The livestock industry, the sweetener industry and now the ethanol industry have taken advantage of these low prices by making corn their primary feedstock.

## What's Being Done?

Two major pieces of federal legislation will create incentives for non-corn based biofuel in the U.S. First, the Energy Independence and Security Act of 2007, passed Congress in December 2007, sets a renewable fuel standard of 36 billion gallons by 2022, of which 15 billion gallons can be corn-based ethanol with the remaining coming from so-called "advanced biofuel," which includes cellulosic biofuel and biomass-based biodiesel.

Second, a new Farm Bill was scheduled to be completed in 2007, but has been pushed back to at least early 2008. Both the House and Senate versions of the Farm Bill contain a "Bioenergy Crop Transition Assistance Programme" to create economic incentives for farmers to begin growing cellulosic biofuel feedstocks; a bioenergy repowering programme to provide federal loans and grants for promoting more public and private investment in the construction of cellulosic biorefineries; and another programme to encourage more production of on-farm energy, including biodiesel.

States have also taken a leadership role in pushing for more environmentally friendly biofuel feedstocks. For example, a Minnesota working lands bioenergy programme allows perennial crops to be grown under long-term easements for the bioenergy market. Other states with advanced biofuels legislation include Washington, Oregon and Massachusetts.

## INDIRECT IMPACTS ON BIODIVERSITY

The previous chapter describes the destruction of natural ecosystems and the threat to global biodiversity caused by current biofuel production systems, but the production of biofuel also has other effects with indirect consequences for biodiversity. This chapter focuses on the biofuel impact on two global issues that will have a profound impact on biodiversity—climate change and the availability of clean water.

## ENERGY USE AND CLIMATE CHANGE

Current global biofuel development, supported by governments around the world, is justified in part by the idea that biofuels are carbon-neutral—i.e., that they emit as much carbon dioxide as they take up during plant growth while emitting less carbon dioxide than comparable fossil fuels. But this simplified perception often fails to take into account both the whole life-cycle of biofuels and carbon dioxide ($CO_2$) emissions due to land-use change. Well-managed biofuel production, incorporated into current agricultural production systems, can reduce greenhouse gas emissions.

However, if biofuel production encroaches on native vegetation, it is unlikely that even the most energy-efficient cropping system could offset the load of greenhouse gases emitted during the clearing of land. Global warming will profoundly affect biodiversity worldwide. Habitats for animals and plants change as temperatures increase and as rainfall patterns change. Scientists have estimated that up to 60 percent of northern latitude habitats could be affected by global warming.

Examples abound, such as the extinction of a frog species by a climate-induced fungal disease documented in the Monteverde Cloud Forest Preserve in Costa Rica. Other species will not be able to move fast enough to adapt to changing habitat and climate. Biofuel production involves the use of fossil fuel energy on the farm, in transport and in the conversion process. Energy is needed for land tillage, pest control, harvesting, drying and storage, as well as to operate the fermentation plant.

Land-use change also contributes significant $CO_2$ emissions. In Brazil, for instance, 80 percent of greenhouse gas emissions in the country come from deforestation, and this is at least in part a consequence of the expansion of sugarcane for biofuel, which replaces soya fields and pushes them into the forests. In Indonesia, $CO_2$ emissions from deforestation account for 83 percent of the country's annual greenhouse gas emissions.

In addition, with most of the highland forests destroyed, many palm oil companies are now moving into Indonesia's lowland tropical forests growing on peaty soils. Because the soils are very wet and poor in oxygen, decomposition is usually very slow, but draining the soils for palm oil plantations is accelerating this decomposition, and carbon that has accumulated for thousands of years is being released. As mentioned earlier, Indonesia recently became the third largest emitter of $CO_2$ worldwide.

The effects of deforestation on the world's climate could be disastrous: not only will massive amounts of carbon dioxide be released, but also any hope of major carbon storage in this unique ecosystem will be lost. The Amazon is a major driver of climate in South America and perhaps worldwide. Costa et al., using the climate change model CCM3, investigated the role of soybean expansion in the onset of Amazon drought, as compared to pastureland extension.

Their results indicated that due to the very high albedo (reflectivity) of the soybean, precipitation was significantly decreased compared to pastureland. This has reduced rainfall by a factor of four compared to the time before clearing. Nitrogen also plays a role in accelerating global warming. Some of the fertilizer in the form of nitrogen will eventually be released into the atmosphere as nitrous oxide ($N_2O$), a greenhouse gas 296 times more potent than $CO_2$.

Excess nitrogen also has direct negative impacts on the biodiversity of terrestrial, marine and freshwater ecosystems. Terrestrial plant species such as prairie grasses adapted to low nitrogen levels are endangered and eutrophication in fresh water is common, with hypoxic zones developing around the world. A recent study by the National Academy of Sciences also suggests that nitrogen in rain is accelerating emissions of $CO_2$ from peat bogs, rendering them carbon sources instead of sinks.

**Water use and Water Quality**

The depletion of water supplies is a global challenge to biodiversity, and biofuel production has contributed to additional agricultural water consumption. The growing, harvesting and processing of biofuel feedstocks in most cases withdraws considerable water. For example, counting only the water used in refining, corn-based ethanol refineries in the U.S. use about four gallons of water for every gallon of ethanol produced.

A typical 100 million gallon per year ethanol plant which is currently the norm in the Midwest U.S., would use about 400 million gallons of water annually—more than a million gallons per day. This is equivalent to the water needs of a community of 10,000.

**Table. Water Consumption by Biofuel for Transportation**

| Fuel type/process | Water use per unit energy [gal/MMBTU] | Water consumption per unit fuel Gal water/gal fuel |
|---|---|---|
| Conventional oil and gas refining | 7-20 | ~1.5 |
| Grain ethanol processing | 12-160 | ~4 |
| Corn irrigation EtOH | 2500-31600 | ~980 |
| Biodiesel processing | 4-5 | ~1 |
| Soybean irrigation biodiesel | 13800-60000 | ~6500 |
| Lignocellulose EtOH | 24-150 | ~2-6 |
| Lignocellulose diesel | 14-90 | ~2-6 |

**Tab]le. Biofuel Water Use Intensity for Selected Feedstocks and Processes.**

| Fuel Type and Conversion Process | Feedstock | Process Water Use gal water/gal fuel | Process Water Consumption gal water/gal fuel | Feedstock Water Demand acre-ft/acre | Feedstock Water Consumption (irrigated) gal water/gal fuel | Biofuel Yield gal fuel/acre |
|---|---|---|---|---|---|---|
| Ethanol | Corn | 2-6 | 4 | 1.2 | 980 | 400 |
| Starch | Sorghum | 2-6 | 4 | 1.0 | 1900 | 170 |
| Sugar | Sugar cane | 2-6 | 4 | 2.0 | 1160 | 560 |
| Sugar | Sugar beets | 2-6 | 4 | 2.3 | 1360 | 550 |
| Ethanol | Switchgrass | ~3-12 | ~2-6 | ~2,3 | Rain fed | 500-800 |
| Cellulose | Woody | ~3-12 | ~2-6 | ~2,5 | Rain fed | 500-800 |
| Biodiesel from | biomass | 0.3-3 | ~1 | ~0.8 | 6500 | 40 |
| oil extraction | Soybeans | 0.3-3 | 1 | ~1.5 | 6100 | 80 |
| Transesterification | Sunflower | 0.3-3 | ~1 | 2,5 | Rain fed | 510 |
| Transesterification | Oil Palm | 0.3-3 | ~1 | Unknown | Unknown | 3,000-15,000 |
| Transesterification | Algae | | | | | |

Water supplies may also be contaminated by pesticides, nutrients and sediment during crop growth, as well as by byproducts of biofuel processing. For example, Palm Oil Mill Effluent (POME), the residue that remains when palms are crushed, contains water, fat residues and fibrous material from the palm, and is high in carbon and low in nitrogen.

When it runs off into water bodies, its high organic carbon content and turbidity can cause organic pollution and reduce the oxygen available for aquatic life. For every ton of fresh palm fruit that is processed, approximately 0.65 tons of POME are produced. Currently in Malaysia POME is the largest pollutant discharge into rivers. Corn ethanol production also produces significant organic waste, which must be treated before the waste water is released into the environment.

### Indirect Land Use Change

A potentially significant, but difficult to quantify, effect of biofuel production on biodiversity arises as the price of corn increases and more acreage is switched from other crops to corn. In the Midwestern U.S., where much of the country's corn is grown, sown acreage of soy has decreased as farmers have expanded corn plantings, and this has contributed to a near doubling in the price of soy. This in turn has helped spur more rapid expansion of soy plantation in Brazil, much of it on formerly forested areas of Amazonia and the wooded savannahs of the Cerrado region.

It would be too simplistic to suggest a one to one correspondence between land taken out of soy production in the U.S. and rainforest converted to soy in Brazil. Several other factors have contributed to the rising demand for soy worldwide, including changing diets in the developing world and the high price of oil. And the phenomenon of tropical forest conversion being driven by international demand for farm products is nothing new. Nevertheless, it is a cruel irony that even the promotion of more sustainable land uses in the North for biofuel feedstocks (such as restoration of mixed perennial grasses in the Corn Belt) may indirectly contribute to biodiversity loss half a world away.

## FUTURE OPPORTUNITIES: CELLULOSIC ETHANOL

The impact of current biofuel production systems on biodiversity is significant. However, if done sustainably, and if combined with a larger strategy of energy conservation and a diversity of renewable energy sources, biofuel does provide an opportunity to diversify cropping systems and land use, and benefit biodiversity. One of the most promising directions for biofuel production is the potential of cellulosic ethanol. Cellulosic ethanol can be produced from hemicelluloses and cellulose in plants and plant residues, and feedstocks can include both dedicated energy crops and wastes from agriculture, industry or forestry.

Cellulosic ethanol is chemically comparable to ethanol produced from starches and sugars. A recent life-cycle analysis revealed that ethanol from switchgrass generally has a slightly better net energy value than corn ethanol and produces significantly lower emissions of greenhouse gases. Another recent study found that switchgrass produced 540 percent more energy than needed to grow, harvest and process it into cellulosic ethanol. Because cellulosic ethanol production allows the use of the entire plant rather than just the starch or sugar, dedicated energy crops offer considerably higher yields.

They also require fewer fossil fuel-derived inputs, such as fertilizer and pesticides, and sequester carbon through a deep root system. Besides the energy-efficiency and greenhouse gas benefits, cellulosic ethanol production also provides farmers with an opportunity to diversify cropping systems. Growing a mix of perennial grasses and shrubs additionally offers environmental advantages over monoculture crop production, as their large root systems stabilize the soils, sequester carbon, regulate water runoff, attract wildlife and support biodiversity.

As a recent report by Blann points out, properly managed agricultural lands can serve as buffers for wildlife, providing habitat that is more compatible with biodiversity. These working lands would be compatible with the conservation of natural communities and as much as possible recognize the structure and function of ecosystems that had initially been displaced by intensive agriculture. If grasses, particularly native swards, become a major feedstock for ethanol plants, there is a strong possibility of developing landscapes that will support higher levels of biodiversity than at present.

The challenge of cellulosic ethanol production lies in efficiently fermenting the sugars found in hemicelluloses and cellulose. This depends on the right types of yeasts and enzymes, and it is estimated that it will still be at least several years before the enzymes will be available on a commercial scale. However, some demonstration cellulosic ethanol plants are currently under construction in the U.S. and elsewhere. If cellulosic ethanol becomes commercially viable, it has the potential to bring dramatic changes to the agricultural landscape.

It could open new lands to production that were previously unsuitable for traditional row crop farming, leading to diverse income sources for some farmers and ranchers. There are other questions that have been raised in connection to cellulosic ethanol development. For example, is there sufficient land available to support cellulosic ethanol production on a large scale? How will cellulosic feedstocks compete with other feedstocks and what will be the consequences for family farmers? Will the cultivation of perennial non-edible feedstocks compete with edible feedstocks? Who will control the patents for all aspects of cellulosic ethanol production and will the benefits be limited to the corporate patent holders? Can cellulosic ethanol development be kept from following the industrial model dictated by much of the current research and development?

It is difficult at this point to answer all these questions, but it is important to recognize that there are many solutions available, including local production of feedstocks, the local ownership of processing plants and sustainable production practices. And national and international policies will play a large role in the industry's future development.

## IMPACT OF BIOFUEL ON BIODIVERSITY

Developing policy recommendations for addressing the impact of biofuel on biodiversity is complicated by the variety of conditions under which biofuel feedstocks are grown, as well as the impact that this industry has on the larger agricultural and energy economies. Little doubt remains that the biofuel industry is now well-established and will maintain a strong presence worldwide for the foreseeable future.

But it is early in the industry's development—the petroleum and coal industries have taken over a century to reach their current level of infrastructure—and there is still time to influence the direction of biofuel development. Ethanol and biodiesel are currently being overlaid on a broken agricultural production system. Many of the biodiversity impacts of biofuel feedstock production are not inherent to biofuel, but are more a symptom of damaging agricultural production systems and policies.

These production systems and policies will need to be changed to create a more sustainable biofuel environment. Importantly, the enormous new demand for biofuel, coupled with the diversity of potential feedstocks, creates a unique opportunity to restructure agricultural production systems. The good news is that a growing number of governments, non-profit organizations and companies are recognizing the need to ensure that global biofuel markets develop in a sustainable manner.

As mentioned earlier in the case studies, the Roundtable on Sustainable Palm Oil is trying to develop standards and guidelines for plantation managers in Malaysia and Indonesia and hopes to have "sustainable" palm oil on the market by the end of 2008. Brazil is attempting to respond to criticism from Europe by establishing environmental and industrial standards certificates for exported ethanol. Europe itself is re-thinking its mandate for biofuel and considering strong environmental criteria for biofuel production.

And the U.S. Energy and Farm Bills include greenhouse gas criteria in the incentives created to help meet a new mandate of 21 billion gallons of non-corn ethanol by 2022. The bad news is that these efforts are thus far very small in the context of the global agricultural system, very late from a biodiversity standpoint, and even if implemented, may by their design have a limited impact. Urgent action is needed. From a biodiversity perspective, we have identified three components of the biofuel industry with major ecological implications that must be addressed immediately.

## LAND-USE

The conversion of native vegetation to agricultural uses is without question the greatest concern with the growth of biofuel production. As documented in the case studies, the decimation of critical ecosystems in Malaysia, Indonesia and Brazil is an ecological disaster that has been exacerbated by new biofuel demand. Any benefits that accrue from sound biofuel production in other parts of the world are more than offset by the loss of these ecosystems.

The challenge, of course, is that stopping biofuel production does not bring this ecological destruction to a halt. Products like palm oil, sugar and soybean have multiple uses, and high petroleum prices create an even greater incentive to find new industrial uses for these products. Additionally, as biomass itself is becoming increasingly valuable, even crop-specific criteria or limits may not have the desired effect on land clearing for agricultural or forest production. Unless land use policies are strengthened and enforced—or the ecological services that these ecosystems provide are more significantly valued—this land-use conversion will continue.

## THE SUSTAINABILITY OF AGRICULTURAL SYSTEMS

Agricultural production is not a monolith, and the ecological services produced from different production systems can vary enormously. The trend in biofuel feedstock use over the past decade has been to expand the use of already abundant agricultural commodities, which generally results in diminished habitat. However, biofuels are not feedstock-specific, so there is an increasing opportunity to introduce new and more sustainable crops and cropping systems to meet some of the need.

If sustainability can be incorporated into this emerging industry, whether through a certification system, policy incentives, or research and development into beneficial and diverse cropping systems, it provides a unique opportunity to get more benefits out of large-scale agricultural production systems.

## COMMODITY PRICES

One of the most contentious issues around the growth in biofuel production is the impact that it has had on commodity prices. Many critics quickly blamed biofuel production for rising food prices, despite the fact that many other factors, from rising petroleum prices to corporate consolidation to climate change, contribute to the rise in global food prices.

Biofuel demand certainly has contributed to the expansion of sugarcane and soybean in Brazil and palm oil in southeast Asia. But these higher prices also create the window of opportunity for the development of preferable feedstocks. This is perhaps most notable in the U.S., where high corn prices have instigated tremendous public and private investment into cellulosic ethanol research and development around perennial grasses that offer multiple ecological benefits.

Biodiversity appears to suffer the most when commodity prices are excessively high or low. When prices are high, agricultural production expands at a frenzied pace. On the flip side, when prices are low, agricultural land rarely goes out of production, but instead the agribusiness industry takes advantage by pumping cheap grains through industrial uses such as livestock factory farms. Moderating these price distortions, which was an integral part of agricultural policy until the mid-1990s, is not a solution by itself, but it does help to alleviate some of these drivers of agricultural expansion.

The complexity of the situation works against any silver bullet solution. Changing inappropriate policies in any particular country is helpful, but could very well result in simply pushing the industry to a different location. To address the multiple drivers in this issue, we provide the following recommendations.

### Protect Native Ecosystems and Indigenous Lands

The most significant biodiversity threat is the potential for biofuel production to extend agriculture's encroachment on native vegetation. Biofuel feedstocks are simply the latest in a long line of commercial crops that have driven forest conversion in the tropics. Lax enforcement of land protection laws in Malaysia, Indonesia and Brazil have all contributed to the proliferation of industrial agricultural production. In the United States, higher commodity prices are convincing farmers to take land out of reserve programs and into production.

Demand for biofuels has increased economic incentives to produce agricultural commodities, while the ecosystem services provided by native vegetation, including protection of genetic resources, are not appropriately valued. Without effective protection of native ecosystems, in combination with incentive programs or other mechanisms that more accurately express the full value of ecosystem services, the biofuel boom will accelerate the encroachment of agriculture onto native lands. Where markets and national governments both fail to protect natural ecosystems from encroachment, indigenous peoples sometimes succeed.

Nepstad, et al have shown that the rate of forest conversion in the Brazilian Amazon is dramatically lower in areas controlled by indigenous peoples, referring to indigenous lands as, "currently the most important barrier to Amazon deforestation." Recognizing and supporting the rights of indigenous peoples is therefore not only a matter of social justice, but also essential to stemming the tide of tropical forest destruction, whether spurred by biofuel markets or other forces.

The growing demand for biofuel, particularly from the EU, has created an economic predicament for many less-developed countries. Developed countries and transnational corporations are willing to invest substantially in biofuel feedstock production, but the market for ecological services has languished. We are still a long way from valuing the benefits of tropical rainforests, which include genetic diversity, habitat, carbon sequestration, water cycling and purification, and countless products of value that are uniquely produced in a rainforest environment.

Specific policies, such as the EU's requirement for the use of biodiesel, have contributed significantly to the destruction of rainforest in Southeast Asia. Hopefully, the EU's promising new steps for environmental protection in the production of biofuel will have a positive effect. Ultimately, it is critical that when the EU and the U.S. use policy to drive the growth of the biofuel industry, the destruction of native ecosystems is discouraged.

### Make Sustainability a Priority for all Biofuel Production

One of the main reasons for broad policy and public support of biofuel has been perceived environmental and rural development benefits. From a biodiversity perspective, biofuel feedstock production provides an opportunity to diversify agricultural cropping systems and generate more environmental benefits from agricultural land, while keeping farmers on the land.

But a more sustainable biofuel production system simply cannot get off the ground if it is competing on the same economic terms as the fossil fuel industry on one side and industrial agriculture on the other. For biofuel to succeed, policies need to assure that sustainability is a priority for all biofuel production.

To that end, policies are needed to encourage more sustainable production of biofuel feedstocks, which could potentially include economic incentives for meeting sustainability criteria, procurement preferences for sustainable biofuel, and greater research and investment in more environmentally beneficial biofuel feedstocks to accelerate the transition to the next generation of biofuel.

Sustainability criteria can take many different paths. Organic foods certification provides an example of the differentiation of a small but growing niche in the food production industry. Standard systems for biofuel that have been initiated include the Roundtable for Sustainable Palm Oil, the Roundtable for Sustainable Soy and the Roundtable for Sustainable Biofuels.

While these approaches may certainly result in better environmental and social practices associated with biofuel feedstock production, there are many concerns about the overall goals, stakeholder participation and effectiveness of such voluntary "market" based approaches.

Considering that the expansion of biofuel is largely a result of mandates and targets set at the state, national and international levels, it seems clear that standards and certification will need policy-level support to be effective.

Only in this manner will such standards actually be able to influence feedstock production in a meaningful way.

## Moderate the Environmental Damage that Results from the Dramatic Price Volatility in Agricultural Commodities

Corn dominates the U.S. biofuel feedstock industry, as well as the industrial livestock feed industry, because no other feedstock could compete against low corn prices. Billions of private, state and federal dollars were invested in using up cheap corn.

Now, even though corn prices have risen substantially, the ethanol and livestock industries remain just as corn-dependent because there has not been adequate research on other environmentally beneficial feedstocks.

These price fluctuations in agriculture are devastating for farmers, environmentally destructive and even have harmful implications on the diet of consumers. Yet since much of the agribusiness industry thrives on market volatility, policies that traditionally assured well-functioning commodity markets have been dismantled.

As stated earlier in this paper, the most productive agricultural systems are welldiversified cropping systems with long crop rotations and several intermixed crops. Industrial production, on the other hand, tends to prefer one crop in the manufacturing process.

This preference for a single crop tends to create enormous markets for particular crops like corn, soybean and palm oil at the expense of other crops that work well in these agricultural systems. This industrial model of agriculture often leaves farmers with little choice but to grow these primary commodities, no matter what happens to commodity prices.

The price variability in agricultural commodity markets is intrinsic to a system that is so dependent on uncontrollable factors such as weather and pest management.

But these vagaries have been exacerbated by an agribusiness industry that thrives on price fluctuations, and has contributed to agriculture's continued encroachment on native lands. Extreme price volatility makes it more challenging for new feedstocks to enter an unstable biofuel market.

The U.S. Farm Bill used to have a series of tools in place to manage supply and prices of primary farm commodities. The University of Tennessee's Agriculture Policy Analysis Centre has documented how an updated supply management would work to stabilize market prices. Maintaining functional markets is critical not only for limiting this encroachment but also for creating diversified agricultural systems on land that is farmed.

## Take Advantage of this Rare Opportunity to Redesign the Agricultural and Energy Sectors

A number of factors—from high gasoline prices to Mideast conflicts to E. coli and Mad Cow outbreaks—have converged to create an overall sense of concern about the direction of agriculture and energy production. This disgruntlement has resulted in explosive growth in local foods, hybrid cars and small wind turbines as consumers seek positive alternatives.

Because biofuel can be produced from a variety of plant materials in nearly every inhabited part of the world, the industry is well-suited for local production, thereby reducing the environmental costs of transportation and allowing local communities to benefit from the sustainable production of biofuel feedstocks. Unfortunately, the environmental and economic benefits of local production and ownership have largely been abandoned in favour of huge production facilities focused on export to other regions and countries.

In the Midwestern U.S. state of Minnesota, policies initiated in the 1980s contributed to development of an ethanol industry that was truly homegrown; state incentives favored ethanol plants that were small and cooperatively owned by farmers.

These plants had minimal impact on cropping systems and water supplies. Now, on the other hand, ethanol plants are most likely not locally owned, the production capacity is several times larger, and water availability, air and water contamination, and growth in monocultural corn production has become much more of a concern.

Biofuel presents an interesting dichotomy for those concerned about the environmental performance of agriculture; while the current industry has contributed to harmful, monocultural production practices, a future biofuel industry based on perennial feedstocks provides one of the only large-scale opportunities to diversify agricultural production practices.

While many environmental groups have correctly pointed out the looming ecological catastrophe from unchecked growth in palm oil and soy plantations, they often unfairly paint the entire concept of biofuel as inherently unsustainable.

Admittedly, creating a truly sustainable biofuel industry is no easy task. This once-in-a-generation opportunity to create a renewable fuel industry that is not in the pocket of the fossil fuel or corporate agribusiness industries will become increasingly difficult as investments are made in the wrong kind of biofuel. Finally, the environmental risks and opportunities associated with biofuel demonstrate the crucial role that federal and international policy must play in this arena.

With a fossil fuel industry that has thrived for decades on outrageous subsidies, tax breaks and military interventions, liquid fuels simply do not have a functional market for "free market" solutions to work.

And with the enormous investment from the agribusiness industry on the production and use of a very few energy-intensive crops, agricultural commodity markets are also severely distorted at the expense of environmentally beneficial cropping systems.

Some biofuel policies, most notably the EU's incentive for biodiesel consumption, were likely crafted with good intentions but have backfired and exacerbated environmental issues in other parts of the world. Reforming this incentive would provide some immediate environmental benefit, but is unlikely to slow the long-term investment in industrially produced biofuel. A biofuel industry that supports an ecologically beneficial agriculture will flounder without public investment in appropriately targeted incentives and research and development.

In other words, simply stopping bad policies is not enough; instead, a specific vision, coupled with strategic policy development, is needed to craft a biodiversity-supporting biofuel industry.

## Chapter 6

# The Economics of Ecosystems and Biodiversity

Nature provides human society with a vast diversity of benefits such as food, fibres, clean water, healthy soil and carbon capture and many more. Though our well-being is totally dependent upon the continued flow of these "ecosystem services", they are predominantly public goods with no markets and no prices, so are rarely detected by our current economic compass.

As a result, biodiversity is declining, our ecosystems are being continuously degraded and we, in turn, are suffering the consequences. Taking inspiration from ideas developed in the Millennium Ecosystem Assessment, our initiative, The Economics of Ecosystems and Biodiversity (TEEB), aims to promote a better understanding of the true economic value of ecosystem services and to offer economic tools that take proper account of this value.

We are confident that the results of our work will contribute to more effective policies for biodiversity protection and for achieving the objectives of the Convention on Biological Diversity. TEEB is in two phases and this interim report summarizes the results of Phase I. It demonstrates the huge significance of ecosystems and biodiversity and the threats to human welfare if no action is taken to reverse current damage and losses. Phase II will expand on this and show how to use this knowledge to design the right tools and policies.

The world has already lost much of its biodiversity. Recent pressure on commodity and food prices shows the consequences of this loss to society. Urgent remedial action is essential because species loss and ecosystem degradation are inextricably linked to human well-being. Economic growth and the conversion of natural ecosystems to agricultural production will, of course, continue. We cannot – and should not – put a brake on the legitimate aspirations of countries and individuals for economic development.

However, it is essential to ensure that such development takes proper account of the real value of natural ecosystems. This is central to both economic and environmental management. If we do not adopt the right policies, the current decline in biodiversity and the related loss of ecosystem services will continue and in some cases even accelerate – some ecosystems are likely to be damaged beyond repair. Findings on the cost of inaction suggest that, with a "business-as-usual" scenario, by 2050 we will be faced with serious consequences:

- 11% of the natural areas remaining in 2000 could be lost, chiefly as a result of conversion for agriculture, the expansion of infrastructure, and climate change.
- Almost 40% of the land currently under low-impact forms of agriculture could be converted to intensive agricultural use, with further biodiversity losses.
- 60% of coral reefs could be lost – even by 2030 – through fishing, pollution, diseases, invasive alien species and coral bleaching due to climate change.

Current trends on land and in the oceans demonstrate the severe dangers that biodiversity loss poses to human health and welfare. Climate change is exacerbating this problem. And again, as with climate change, it is the world's poor who are most at risk from the continuing loss of biodiversity. They are the ones most reliant on the ecosystem services which are being undermined by flawed economic analysis and policy mistakes. The ultimate aim of our work is to provide policy makers with the tools they need to incorporate the true value of ecosystem services into their decisions.

Since ecosystem economics is still a developing discipline – we describe the key challenges in developing and applying suitable methodologies. In particular, there are ethical choices to be made between present and future generations and between peoples in different parts of the world and at different stages of development. Without taking these aspects into account, the Millennium Development Goals cannot be achieved. Some promising policies are already being tried out.

We describe several that are already working in some countries and could be scaled up and/or replicated elsewhere. These examples come from many different fields, but they convey some common messages for developing the economics of ecosystems and biodiversity:

- Rethink today's subsidies to reflect tomorrow's priorities.
- Reward currently unrecognized ecosystem services and make sure that the costs of ecosystem damage are accounted for, by creating new markets and promoting appropriate policy instruments.

- Share the benefits of conservation.
- Measure the costs and benefits of ecosystem services.

The economic approach we will be working on in Phase II will be spatially specific and will build on our knowledge of how ecosystems function and deliver services. We will also examine how ecosystems and their associated services are likely to respond to particular policy actions. It will be essential to take account of the ethical issues and equity, and of the risks and uncertainty inherent in natural processes and human behaviour. Most biodiversity and ecosystem benefits are public goods that have no price.

There are different approaches for solving this problem. Notably, we can adopt policies that reward preservation of the flow of these public goods, or we can encourage "compliance markets" which attach tradable values to the supply or use of these services. One example is payments for ecosystem services (PES). These can create demand so as to correct the imbalances which harm biodiversity and impede sustainable development. Phase II will examine the investment case for PES, but also for other new and innovative instruments.

New markets are already forming which support and reward biodiversity and ecosystem services. To be successful, they need the appropriate institutional infrastructure, incentives, financing and governance: in short, investment and resources. In the past, the state was often considered solely responsible for managing ecosystems. Now it is clear that markets can also play their part – often without drawing on public money. The fundamental requirement is to develop an economic yardstick that is more effective than GDP for assessing the performance of an economy.

National accounting systems need to be more inclusive in order to measure the significant human welfare benefits that ecosystems and biodiversity provide. By no longer ignoring these benefits, such systems would help policy makers adopt the right measures and design appropriate financing mechanisms for conservation. Countries, companies and individuals need to understand the real costs of using the Earth's natural capital and the consequences that policies and actions, individual or collective, have on the resilience and sustainability of natural ecosystems.

We believe that policies which better reflect the true value of biodiversity and natural ecosystems will contribute to sustainable development by helping to secure the delivery of ecosystem goods and services, particularly food and water, in a transparent and socially equitable way. This will not only protect biodiversity, ecosystems and the associated ecosystem services, but will also improve the well-being of our present generation and the generations to come. If we are to achieve our highly ambitious goals we will need to draw on the knowledge, skills, and talent of countries, international bodies, academia, business and civil society from around the world. We look forward to working together openly, flexibly and constructively and to seeing further substantive progress in 2009 and 2010.

## BIODIVERSITY AND ECOSYSTEMS TODAY

The connection between nature, its preservation and destruction, human welfare, and finally, money. Historically, nature's role as the nurturer of human society was accepted as a given, and the "maternal" image of nature abounds in rituals, epics and beliefs across all societies and times. Over the last half century however, the intricate relationship between human wealth and welfare and biodiversity, ecosystems and their services is increasingly being understood in ecological and economic terms.

Our knowledge, of the many dimensions of this relationship is improving fast. At the same time, we are recognizing increasing natural losses – worsening environments, declining species. Many high-profile species such as pandas, rhinos and tigers face extinction, while rainforests, wetlands, coral reefs and other ecosystems are under huge pressure from human activity. Natural disasters such as floods, droughts and landslides are today almost commonplace, while food and water shortages have recently been commanding world attention.

While there is some understanding that these many phenomena are in some way connected, there is at the same time an expectation that "normal service" will soon be resumed. There seems to be little appreciation of the many dimensions of biodiversity loss, or the connections between biodiversity loss, climate change and economic development.

Species loss and ecosystem degradation are inextricably linked to human well-being, and unless we take urgent remedial action, "normal service" – in the sense of being able to enjoy the benefits that our environment affords us – may never be resumed. Humanity receives countless benefits from the natural environment in the form of goods and services (generally grouped under the collective title of ecosystem services) such as food, wood, clean water, energy, protection from floods and soil erosion.

Natural ecosystems are also the source of many life-saving drugs as well as providing sinks for our wastes, including carbon. Human development has also been shaped by the environment, and this inter linkage has strong social, cultural and aesthetic importance. The well-being of every human population in the world is fundamentally and directly dependent on ecosystem services. However, the levels of many of the benefits we derive from the environment have plunged over the past 50 years as biodiversity has fallen dramatically across the globe. Here are some examples:

- In the last 300 years, the global forest area has shrunk by approximately 40%. Forests have completely disappeared in 25 countries, and another 29 countries have lost more than 90% of their forest cover. The decline continues.
- Since 1900, the world has lost about 50% of its wetlands. While much of this occurred in northern countries during the first 50 years of the 20th century, there has been increasing pressure since the 1950s for conversion of tropical and sub-tropical wetlands to alternative land use.
- Some 30%of coral reefs – which frequently have even higher levels of biodiversity than tropical forests – have been seriously damaged through fishing, pollution, disease and coral bleaching.
- In the past two decades, 35% of mangroves have disappeared. Some countries have lost up to 80% through conversion for aquaculture, overexploitation and storms.
- The human-caused (anthropogenic) rate of species extinction is estimated to be 1,000 times more rapid than the "natural" rate of extinction typical of Earth's long-term history.

The effect of trends such as these is that approximately 60% of the Earth's ecosystem services that have been examined have been degraded in the last 50 years, with human impacts being the root cause. Further declines are projected over the coming decades because of factors such as population growth, changing land use, economic expansion and global climate change. Leading international economic organizations such as the World Bank and the Organisation for Economic Co-operation and Development (OECD) confirm these worrying predictions.

The OECD has described a highly daunting combination of challenges facing humanity: tackling climate change, halting biodiversity loss, ensuring clean water and adequate sanitation, and reducing the human health impacts of environmental degradation.

The pressures have intensified even in the short time since the publication of the Millennium Ecosystem Assessments in 2005. In 2007, more people were living in urban than rural areas for the first time in human history.

During 2007 and 2008, the push to develop bio-fuels resulted in massive changes in land use and a steep increase in the price of some staple food crops. Continuing high rates of economic growth in some of the large developing economies have resulted in demand outstripping supply for several commodities, putting even greater pressure on natural systems.

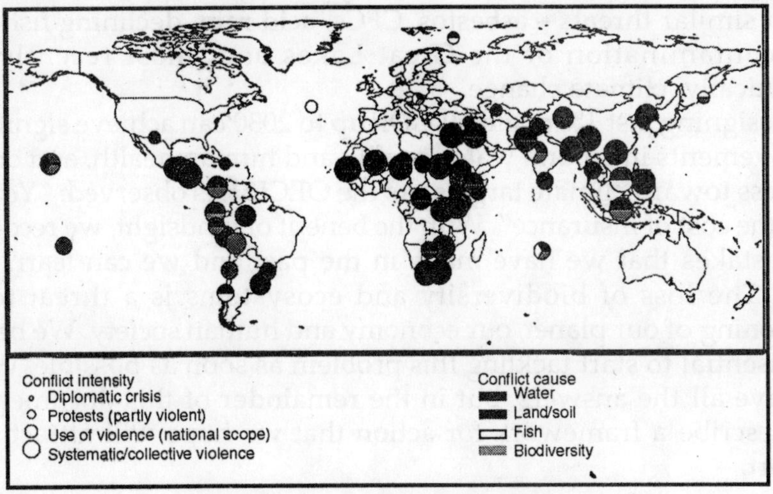


**Fig.** The Earth's Ecosystem Services

Recent evidence of climate change suggests much faster and deeper impacts than previously predicted, including the risk of human conflicts caused by competition for biodiversity resources and ecosystem services. Such trends may change our relationship with nature but not our reliance on it. Natural resources, and the ecosystems that provide them, underpin our economic activity, our quality of life and our social cohesion. But the way we organize our economies does not give sufficient recognition to the dependent nature of this relationship – there are no economies without environments, but there are environments without economies.

There have been many attempts to fill this gap by putting some kind of monetary value on ecosystem services. Such approaches can be helpful, but above all we need to regain a sense of humility about the natural world. As traditional peoples have long understood, we must ultimately answer to nature, for the simple reason that nature has limits and rules of its own. We are consuming the world's biodiversity and ecosystems at an unsustainable rate and this is already starting to have serious socio-economic impacts. If we are to find solutions to the problems we face, we need to understand what is happening to biodiversity and ecosystems and how these changes affect the goods and services they provide.

We then need to look at the way we can use economic tools to ensure that future generations can continue to enjoy the benefits of these goods and services. This is a highly complex challenge, but one which must be met. However, lessons from the last 100 years demonstrate that mankind has usually acted too little and too late in face of similar threats – asbestos, CFCs, acid rain, declining fisheries, BSE, contamination of the Great Lakes and, most recently and dramatically, climate change.

Assigning just 1% of global GDP up to 2030 can achieve significant improvements in air and water quality and human health, and ensure progress toward climate targets. As the OECD has observed: "You can call it the cost of insurance". With the benefit of hindsight, we recognize the mistakes that we have made in the past and we can learn from them. The loss of biodiversity and ecosystems is a threat to the functioning of our planet, our economy and human society. We believe it is essential to start tackling this problem as soon as possible. We do not have all the answers, but in the remainder of this document we will describe a framework for action that we hope will attract wide support.

## BIODIVERSITY, ECOSYSTEMS AND HUMAN WELFARE

The UN Secretary-General's resolute optimism with regard to tackling climate change could also be taken as an appropriate rallying call for addressing the problem of biodiversity loss.

It will indeed take a global response and a concerted effort from all nations and across all sections of society if we are to achieve our goal. Today's global consumption and production patterns are underpinned by ecosystems around the world. Many different types of policy can affect the resilience of natural as well as human-modified ecosystems.

From transport to energy, agriculture to cultural well-being, policies and actions can have many unintended consequences. As demonstrated by the Millennium Ecosystem Assessment, the impacts of cumulative pressures on ecosystems may not be felt for many years, until some tipping points are reached leading to rapid non-linear changes. We begin this chapter with selected examples that illustrate the wide range of effects, from food to health. Then we set out some common themes, especially the disproportionate impact on the poor.

This chapter shows that the implications of ecosystem degradation can be far-reaching, for example the threat to healthcare from the loss of plant species. The result, as this chapter concludes, is that business-as-usual is not an option, even in the short-term.

## PRESSURES ON BIODIVERSITY WILL CONTINUE AND HUMAN WELL-BEING WILL BE AFFECTED

### Food is News on Land

Rising food prices have provoked protests in many countries. In February 2007, tens of thousands of people marched through the streets of Mexico City, demonstrating against a 400% increase in the cost of corn used to make tortillas – blamed on increased demand for bio-fuels in the United States of America. In Asia, many governments had to intervene to ease rocketing rice prices and to manage supplies, while the Philippines also distributed food aid to affected people in rural areas.

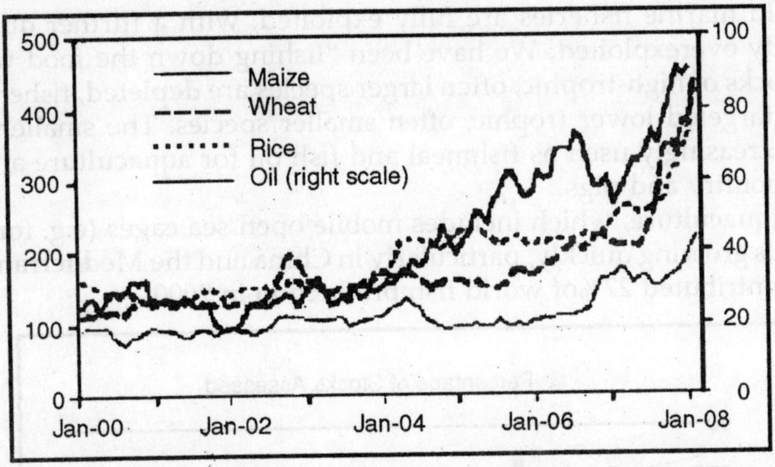


Fig. World Commodity Prices, January 2000-February 2008 (US$/tonne).

There are many causes for the increase in food prices. They include rising demand for food and especially meat (which requires more land per calorie), the rising price of energy (which is an important input) and increasing demand for bio-fuels. In 2007, the food price index calculated by the Food and Agriculture Organization of the United Nations (FAO) rose by nearly 40%, compared with 9% the previous year.

In the first months of 2008 prices again increased drastically. Nearly every agricultural commodity is part of this rising price trend. As demand for basic commodities increases, this raises the pressure to convert natural ecosystems into farmland and to increase the intensity of production from already converted land. Already, the shift toward higher meat consumption is one of the most important causes of deforestation worldwide.

There is no sign that this pressure for conversion from natural ecosystems towards arable land will abate. Demand for food is set to increase as populations grow and their consumption shifts towards more meat.

Supply cannot keep pace as yields are growing only slowly. On top of this, scientists of the Intergovernmental Panel on Climate Change (IPCC) predict in their 2007 report that even slight global warming would decrease agricultural productivity in tropical and subtropical countries.

## AT SEA

More than a billion people rely on fisheries as their main or sole source of animal protein, especially in developing countries. But half of wild marine fisheries are fully exploited, with a further quarter already overexploited. We have been "fishing down the food web". As stocks of high-trophic, often larger species are depleted, fishermen have targeted lower trophic, often smaller species. The smaller fish are increasingly used as fishmeal and fish oil for aquaculture and to feed poultry and pigs.

Aquaculture, which includes mobile open sea cages (e.g. for red tuna) is growing quickly, particularly in China and the Mediterranean, and contributed 27%of world fish production in 2000.

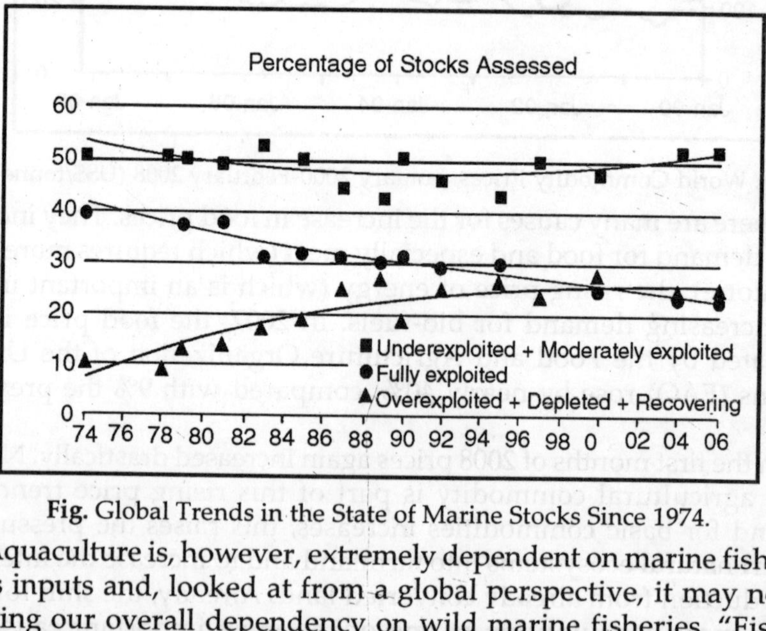


Fig. Global Trends in the State of Marine Stocks Since 1974.

Aquaculture is, however, extremely dependent on marine fisheries for its inputs and, looked at from a global perspective, it may not be reducing our overall dependency on wild marine fisheries. "Fishing down the food web" leads to diverse impacts on the biodiversity of the oceans. The blooms of jellyfish that have increased rapidly worldwide in the last decade are believed to result in part from this situation. Jellyfish have replaced fish as the dominant planktivores in several areas, and there is some concern that these community shifts may not be easily reversible, since the jellyfish also eat the eggs of their fish competitors.

This loss of biodiversity could have disastrous effects on the supply of seafood to the human population and on the economy. There is increasing evidence that species diversity is important for marine fisheries, both in the short term, by increasing productivity, and in the long term, by increasing resilience, while genetic diversity is important particularly for the latter.

A 2006 study concluded that all of the world's commercial fisheries are likely to have collapsed in less than 50 years unless current trends are reversed. It found that low diversity is associated with lower fishery productivity, more frequent "collapses", and a lower tendency to recover after over fishing than naturally species rich systems. The security value of biodiversity can be compared with financial markets.

A diverse portfolio of species stocks, as with business stocks, can provide a buffer against fluctuations in the environment (or market) that cause declines in individual stocks. This stabilizing effect of a "biodiverse" portfolio is likely to be especially important as environmental change accelerates with global warming and other human impacts.

## WATER SUPPLY INCREASINGLY AT RISK

There is also growing pressure on water resources – both the supply of water and its quality. Many parts of the world already live with water stress. The risk of water wars was a major theme at the 2008 World Economic Forum in Davos. The United Nations believes there is enough to go round – but only if we keep it clean, use it wisely and share it fairly. In Asia, the water vital for the irrigation of the grain crops that feed China and India is at risk of drying up because of climate change.

Global warming melts the glaciers that feed Asia's biggest rivers in the dry season – precisely the period when water is needed most to irrigate the crops on which hundreds of millions of people depend. In this example, climate change could accentuate the problems of chronic water shortage and drive the ecosystem service that provides a reliable supply of clean water beyond breaking point.

In many areas, ecosystems provide vital regulating functions. Forests and wetlands can play an important role in determining levels of rainfall (at a regional and local level), the ability of land to absorb or retain that water and its quality when used. In other words, ecosystems play a part in determining whether we have droughts, floods and water fit to drink. The value of this role is often forgotten until it is lost.

### Biofuels Generate much Debate

Bio-energy can play an important role in combating climate change, specifically if biomass is used for heat and electricity generation. However, bio-fuels also are another source of competition for scarce land, and the scale of potential land conversion for agro-fuels is extraordinary.

The International Monetary Fund reports that "although bio-fuels still account for only 1.5% of the global liquid fuels supply, they accounted for almost half of the increase in consumption of major food crops in 2006-2007, mostly because of corn-based ethanol produced in the US". Reports indicate that this pattern could be replicated elsewhere in the world.

### Coral Reefs

Coral reefs are the most biodiversity-rich ecosystems (in species per unit area) in the world, more diverse even than tropical forests. Their health and resilience are in decline because of over fishing, pollution, disease and climate change. Caribbean coral reefs have been reduced by 80%in three decades. As a direct result, revenues from dive tourism(close to 20%of total tourism revenue) have declined and are predicted to lose up to US$ 300 million per year. That is more than twice as much as losses in the heavily impacted fisheries sector.

The underlying explanation for this situation is that in 1983, following several centuries of over fishing of herbivores, there was a sudden switch from coral to algal domination of Jamaican reef systems. This left the control of algal cover almost entirely to a single species of sea urchin, whose populations collapsed when exposed to a species-specific pathogen. When the sea urchin population collapsed, the reefs shifted (apparently irreversibly) to a new state with little capacity to support fisheries. This is an excellent example of the insurance value in biologically diverse ecosystems. The reduction in herbivore diversity had no immediate effect until the sea urchin population plummeted, illustrating how vulnerable the system had become due to its dependence on a single species.

## OUR HEALTH IS AT STAKE

People have known the medicinal value of certain plants for

thousands of years and biodiversity has helped our understanding of the human body. So ecosystems provide huge health benefits, and thus economic benefits. The corollary is that losing biodiversity incurs potentially huge costs, and our knowledge of these is growing. There are significant direct links between biodiversity and modern healthcare:

- Approximately half of synthetic drugs have a natural origin, including 10 of the 25 highest selling drugs in the United States of America.Of all the anti-cancer drugs available, 42% are natural and 34% semi-natural.
- In China, over 5,000 of the 30,000 recorded higher plant species are used for therapeutic purposes.
- Three quarters of the world's population depend on natural traditional remedies.
- The turnover for drugs derived from genetic resources was between US$ 75 billion and US$ 150 billion in the United States of America in 1997.
- The gingko tree led to the discovery of substances which are highly effective against cardiovascular diseases, accounting for a turnover of US$ 360 million per year.

Despite the enormous health benefits, plants are disappearing fast and will continue to do so unless urgent action is taken. The *2007 IUCN Red List of Threatened Species* identified a significant increase in species under threat during this decade. It estimates that 70%of the world's plants are in jeopardy. A recent global study reveals that hundreds of medicinal plant species, whose naturally occurring chemicals make up the basis of over 50%of all prescription drugs, are threatened with extinction. This prompted experts to call for action to "secure the future of global healthcare".

The biodiversity-healthcare relationship also has a strong distributional equity dimension. There is often a mismatch between the regions where benefits are produced, where their value is enjoyed, and where the opportunity costs for their conservation are borne. So the plant species that are the sources of many new drugs are likely to be found in poorer tropical regions of the world. The people that benefit are more likely to be found in rich countries where the resulting drugs are more readily available and affordable. People in these countries therefore have a great incentive to conserve natural habitats in biodiversity-rich parts of the world.

However, such conservation has costs for local people in these parts, in particular the opportunity costs such as the loss in potential agriculture returns of not converting such habitats. Transferring some of the rich world benefits back to local people could be one approach to improving incentives to conserve those natural habitats and species locally that clearly have wider benefits globally. It is clear that if we undermine the natural functions that hold this planet together, we may be creating conditions that will make life increasingly difficult for generations to come – and impossible for those already on the margins of survival.

## GROWTH AND DEVELOPMENT

Population growth, increasing wealth and changing consumption patterns underlie many of the trends we have described. Unsustainable resource use has been evident in the developed world for many years. The ecological footprints of Europe, the United States of America and Japan are much higher than those of developing countries. And the emerging economies are catching up. India and China both have ecological footprints twice the size of their "bio-capacities" – the extent to which their ecosystems can generate a sustainable supply of renewable resources.

Brazil, on the other hand, has one of the world's highest "bio-capacities", nearly five times as large as its ecological footprint, yet this is declining as a result of deforestation. Under current practices, meeting the food needs of growing and increasingly affluent populations will further threaten biodiversity and ecosystem services. Based on population projections alone, 50% more food than is currently produced will be required to feed the global population by 2050.

Irrigated crop production will need to increase by 80% by 2030 to match demand. Already, 35% of the Earth's surface has been converted for agriculture, limiting scope for the future productivity of natural systems. The livestock sector already represents the world's single largest human use of land. Grazing land covers 26% of the Earth's surface, while animal feed crops account for about a third of arable land.

Extending agricultural production will have consequences for biodiversity and ecosystem services as more land is converted for food production. The expanding livestock sector will be in direct competition with humans for land, water and other natural resources. Livestock production is the largest sectoral source of water pollutants. It is also a major factor in rising deforestation: 70% of land in the Amazon that was previously forested is now used as pasture, and livestock feed crops cover a large part of the remainder.

## CLIMATE CHANGE AND BIODIVERSITY

Climate change is linked to many of the issues we have presented in this chapter. The El Niño-La Niña cycle in the Pacific Ocean is one prominent example of the vulnerability of biodiversity to climate. A small rise in the sea surface temperature in 1976 and 1998 led to a series of worldwide phenomena, which resulted in 1998 being characterized as "the year the world caught fire". Permanent damage includes:

- Burned forests that will not recover within any meaningful human timescale.
- A rise in the temperature of surface waters of the central western Pacific Ocean from an average of 19°C to 25°C.
- Shifts toward heat-tolerant species living inside corals.
- A northward shift in the jet stream.

These types of complex phenomena show us how vulnerable we are to tipping points beyond those linked directly to increasing temperatures and carbon dioxide levels. Biodiversity losses can also contribute to climate change in many complex ways. There are many examples of how over harvesting or changed land-use patterns have triggered social and economic changes leading to greater reliance on carbon. Draining peat lands results in carbon losses.

But predicted changes to climate could cause accelerated rates of carbon release from the soil, contributing in turn to higher greenhouse gas concentrations in the atmosphere. Under the same climatic conditions, grassland and forests tend to have higher stocks of organic carbon than arable land and are seen as net sinks for carbon. Yet deforestation and intensification of cropland areas are rampant. To take account of these complexities we will need more than energy-based econometric models.

We will need to respond to knowledge about how to adapt and how vulnerabilities might arise from global ecological processes. This will require a much deeper dialogue than we have seen so far between economists, climate scientists and ecologists.

## IMPACTS ON THE POOR

A striking aspect of the consequences of biodiversity loss is their disproportionate but unrecognized impact on the poor. For instance, if climate change resulted in a drought that halved the income of the poorest of the 28 million Ethiopians, this would barely register on the global balance sheet – world GDP would fall by less than 0.003%. The distributional challenge is particularly difficult because those who have largely caused the problems – the rich countries – are not going to suffer the most, at least not in the short term.

The evidence is clear. The consequences of biodiversity loss and ecosystem service degradation – from water to food to fish – are not being shared equitably across the world. The areas of richest biodiversity and ecosystem services are in developing countries where they are relied upon by billions of people to meet their basic needs. Yet subsistence farmers, fishermen, the rural poor and traditional societies face the most serious risks from degradation.

This imbalance is likely to grow. Estimates of the global environmental costs in six major categories, from climate change to over fishing, show that the costs arise overwhelmingly in high- and middle-income countries and are borne by low-income countries.

## Gender, Poverty and Biodiversity in Orissa, India

The impact of the loss of biodiversity, often not very visible, has serious implications for poverty reduction and well-being for women as it severely affects the role of women as forest gatherers. Studies in the tribal regions of Orissa and Chattisgarh, states in India which were once heavily forested, have recorded how deforestation has resulted in loss of livelihoods, in women having to walk four times the distance to collect forest produce and in their inability to access medicinal herbs which have been depleted.

This loss reduces income, increases drudgery and affects physical health. There is also evidence to show that the relative status of women within the family is higher in well-forested villages, where their contribution to the household income is greater than in villages that lack natural resources.

**Table. Ecosystem Services and the Millennium Development Goals: links and Trade-offs.**

| cosystem services | Related MDG | Links with targets | Conflicting outcome | Evaluation |
|---|---|---|---|---|
| Provisioning and regulating services | MDG 1: Eradicate extreme poverty and hunger | Steady daily supplies of water, fuelwood and food: these influence the material minimum standard of the lives of the poor, alleviating poverty and hunger | Greater conflicts over water, exploitation of top soil, coastal and marine resources and the resilience of agri-biodiversity could constitute trade-offs | Strong and direct links: Intervention needs to be receptive to ecosystem services, biodiversity and the resilience of cultivated ecosystems |
| Services from, wetlands and forests | MDG 3: Promote gender equality and empower women | Fuelwood and water: adequate availability and and proximity - would help gender equality by reducing this burden that falls mainly on women | There could be greater extraction of groundwater. The enforcement of land rights for women would, however, ensure the prevention | Indirect link |

| | | | |
|---|---|---|---|
| | | of biodiversity loss to a greater extent | |
| Provisioning (medicinal plants) and regulating services (water) | MDG 5: Improve maternal health | Better availability of clean water and traditional medical services would create enabling conditions | Indirect link |
| Provisioning and regulating services | MDG 6: Combat HıV/AIDS, malaria and other dieases | This would be facilitated by widening the availability of clean water | Indirect link |
| Provisioning services | MDG 8: Develop a Global Partnership for Development | Fair and equitable trade practices and a healthy world economic order would reflect the true cost of export/import from the ecosystem services perspective | Indirect link |
| Provisioning and regulating services | MDG 4: Reduce child mortality | Creating enabling conditions, e.g. through clean water | Indirect link |
| Provisioning and regulating services | MDG 2: Achieve universal primary education | Provisioning services might be affected by expansion of education-related infrastructure (schools and roads) | Weak or unclear link |

The Millennium Development Goals (MDGs) represent the world's ambition to attack poverty. Anecdotal evidence abounds showing that achievement of these goals assumes sound environmental practice and governance. An example that powerfully illustrates this point is that of Haiti, where forest degradation and its consequences have jeopardized water availability and agricultural productivity to the point where hunger and poverty elimination (MDG1) has proved impossible, and have severely affected health and child mortality (MDG4, MDG5 and MDG6), to name some of the MDG linkages. In Table, we map ecosystem services against the MDGs. The extent of linkage is deep and broad, suggesting that there are significant risks to the achievement of all MDGs, and not just MDG7 about environmental sustainability, if the current pace of ecosystem degradation and biodiversity losses continues unchecked.

## BUSINESS-AS-USUAL IS NOT AN OPTION

If nomajor new policymeasures are put in place, past trends of biodiversity and ecosystem service loss will continue. In some cases losses will accelerate. In others the ecosystem will be degraded to such an extent that it will not be possible to repair or recover it. These are some of the likely results of inaction:

- Natural areas will continue to be converted to agricultural land, and will be affected by the expansion of infrastructure and by climate change. By 2050, 7.5 million square kilometres are expected to be lost, or 11% of 2000 levels.
- Land currently under extensive (low-impact) forms of agriculture, which often provides important biodiversity benefits, will be increasingly converted to intensive agricultural use, with further biodiversity losses and with damage to the environment. Almost 40% of land currently under extensive agriculture is expected to be lost by 2050.
- 60% of coral reefs could be lost by 2030 through fishing damage, pollution, disease, invasive alien species and coral bleaching, which is becoming more common with climate change. This risks losing vital breeding grounds as well as valuable sources of revenue to nations.
- Valuable mangrove areas are likely to be converted to use for private gain, often to the detriment of local populations. Important breeding grounds will be lost, as will buffers that protect against storms and tsunamis.
- If current levels of fishing continue, there is the risk of collapse of a series of fisheries. The global collapse of most world fisheries is possible by the second half of the century unless there is an effective policy response – and enforcement.
- As global trade and mobility increase, so do the risks from invasive alien species for food and timber production, infrastructure and health.

Business-as-usual is not an option if we wish to avoid these consequences and to safeguard our natural capital and the well-being of future generations. The cost of insufficient policy action is too great. Some solutions are already visible, however, and economics could play an important part. Although forests are at risk of conversion to agriculture, grazing lands and bio-fuel production, they can play a valuable role as carbon sinks and biodiversity vaults, and this capacity could be recognized by a higher market value.

### The Changing use of Land and Changing Services

Humans have been causing biodiversity loss for centuries. By the year 2000, only about 73%of the original global natural biodiversity was left. The strongest declines have occurred in the temperate and tropical grasslands and forests, where human civilizations first developed.

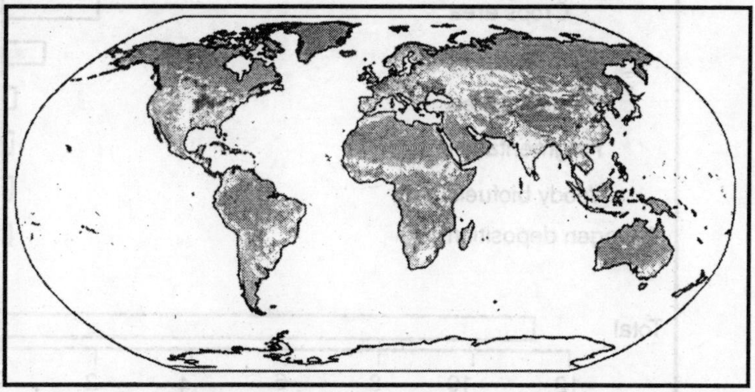

**Fig.** Mean Species Abundance 1970.

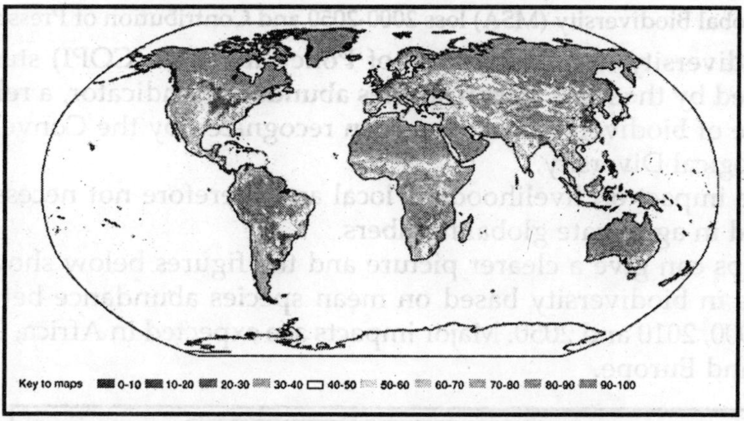


**Fig.** Mean Species Abundance 2000.

A further 11%of land biodiversity is expected to be lost by 2050, but this figure is an average including desert, tundra and polar regions. In some biomes and regions, projected losses are about 20%. Natural areas will continue to be converted to agricultural land, with the ongoing expansion of infrastructure and increasing effects of climate change being additional major contributors to biodiversity loss.

For the world as a whole, the loss of natural areas over the period 2000 to 2050 is projected to be 7.5 million square kilo-metres or around 750million hectares, i.e. the size of Australia. These natural ecosystems are expected to undergo human-dominated land-use change in the next few decades.

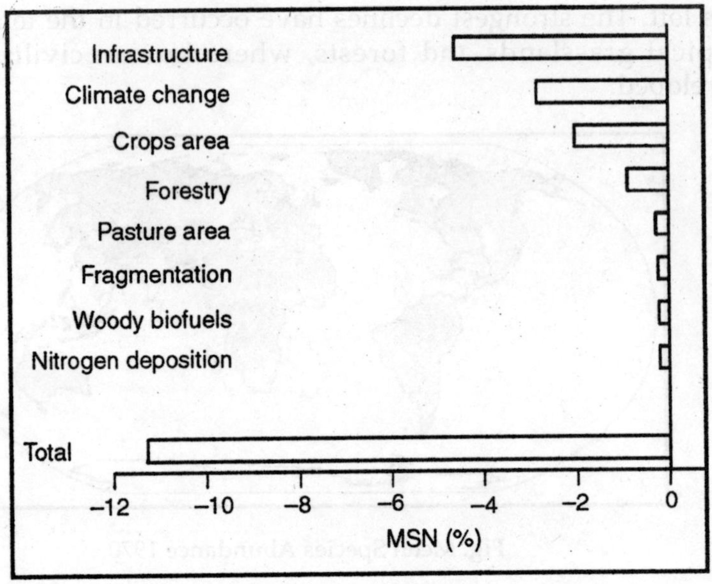


**Fig.** Global Biodiversity (MSA) loss 2000-2050 and Contribution of Pressures.

Biodiversity loss in the Cost of Policy Inaction (COPI) study is measured by the MSA (mean species abundance) indicator, a reliable measure of biodiversity that has been recognized by the Convention on Biological Diversity.

The impact on livelihoods is local and therefore not necessarily reflected in aggregate global numbers.

Maps can give a clearer picture and the figures below show the changes in biodiversity based on mean species abundance between 1970, 2000, 2010 and 2050. Major impacts are expected in Africa, India, China and Europe.

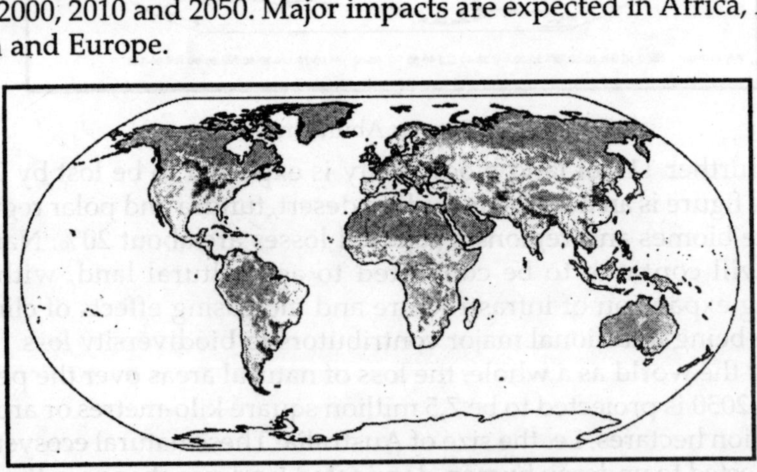

**Fig.** Mean Species Abundance 2010

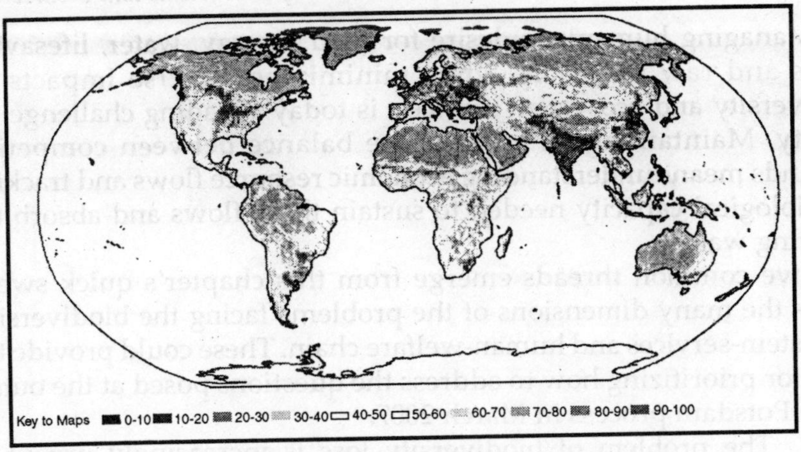


**Fig.** Mean Species Abundance 2050

## Vicious Cycle of Poverty and Environmental Degradation

Haiti is the poorest country in the Western Hemisphere and one of the most environmentally degraded. Over 60% of its income comes as aid from the USA and other countries, and 65% of its people survive on less than $1 a day. Almost all of the country was originally forested but now there is less than 3% cover left. As a consequence, from 1950-1990, the amount of arable land fell by more than two fifths due to soil erosion. At the same time deforestation has diminished evaporation back to the atmosphere over Haiti, and total rainfall in many locations has fallen by as much as 40%, reducing stream flow and irrigation capacity.

The Avezac Irrigation System supports only half of the initially planned 9,500 acres (3,845 hectares). When the rains do come, hillsides no longer efficiently retain or filter water. Due to deforestation, even moderate rains can produce devastating floods. Ground and stream waters are laden with sediment and pollution which has degraded estuary and coastal ecosystems.

As a consequence, nearly 90% of Haitian children are chronically infected with intestinal parasites they acquire from the water they drink. Due to flooding, Haiti has lost half of its hydropower potential since sediment clogged the Peligre Dam. Haiti is a stark example of the "vicious circle" of extreme poverty and environmental degradation.

Much of Haiti's poverty and human suffering derives from the loss of its forests, and extreme poverty is itself one of the root causes of deforestation and a powerful barrier to sustainable forest management. The alleviation of poverty must be a central strategy to restore Haiti's forest and biodiversity.

Managing humanity's desire for food, energy, water, lifesaving drugs and raw materials, while minimizing adverse impacts on biodiversity and ecosystem services, is today's leading challenge for society. Maintaining an appropriate balance between competing demands means understanding economic resource flows and tracking the biological capacity needed to sustain these flows and absorb the resulting waste.

Five common threads emerge from this chapter's quick sweep across the many dimensions of the problems facing the biodiversity, ecosystem-services and human-welfare chain. These could provide the basis for prioritizing how to address the questions posed at the outset of the Potsdam process in March 2007.

- The problem of biodiversity loss is increasingly urgent in terms of the rate and costs of loss and the risks of crossing "tipping points".
- Our growing, if still fragmented, understanding is often sufficient warning to support action.
- We have time to act but that time is fast diminishing.
- Seemingly small changes in one place can have huge though largely unpredictable impacts elsewhere.
- In all cases the poor are bearing the brunt of the situation.

The classic development challenge of increasing economic opportunity and providing goods and services is still with us, but it has now been sharpened by the emerging recognition of global ecological constraints. Similarly, social justice will be threatened if the world continues to deepen the gulf between those who have the use of ecological goods and services and those who do not. Resentment over inequitable use of the planet's resources could erode international collaboration and trust, undermining the benefits of an integrated global economy and even threatening its very existence.

Acting to reduce ecological deficits before being forced to do so is far preferable to the alternative. If we plan reductions by cutting demand for ecological resources, this need not necessarily entail hardship, and may even add growth opportunities to the economy and improve quality of life. On the other hand, as many telling examples from history show, when societies that operate with an ecological deficit experience unplanned reductions in resource use and are forced to rely on their own "bio-capacity", a decline in quality of life, often severe, generally follows.

There is still time to act. A wide variety of strategies and approaches are already being used to drive technological and organizational solutions that reduce human demand on nature. Since the apparent un-sustainability of society's current growth path has often been guided by economic metrics that ignore market and regulatory failures, and accompanied by a policy framework that does not achieve adequate conservation of biodiversity and ecosystems, we must ask two basic questions.

First, what are the economic tools we need to guide us towards a sustainable, ecologically secure future? Second, how can this economics "toolkit" help us to evaluate and reform policies in order to achieve sustainable development, ecological security, and an accompanying level of conservation of ecosystems and biodiversity?

## TOWARDS A VALUATION FRAMEWORK

The previous chapter demonstrated the many dimensions of the continuing decline of ecosystems and biodiversity, its significant human impact and the urgent need for action. Here we consider how the failure to recognize the economic value of wild nature has contributed to this continuing decline. We evaluate the challenges of placing economic values on the benefits of ecosystems and biodiversity that are not currently captured, and consider vital issues of ethics and equity which need to be at the heart of such evaluation.

This chapter identifies the difficulties in evaluating ecosystem services and the main aspects of work we will carry out in Phase II, when we will focus on addressing these difficulties while firming up both a preferred framework and methodologies for estimating ecosystem and biodiversity values.

## MANY FAILURES, ONE PROBLEM

Biodiversity loss and ecosystem degradation continue, despite the fact that policy makers, administrators, NGOs and businesses around the world have been seeking ways to stem the tide. There are many reasons for this, but perverse economic drivers as well as failures in markets, information and policy are significant factors. Markets tend not to assign economic values to the largely public benefits of conservation, while assigning value to the private goods and services the production of which may result in ecosystem damage.

The term market failure can cover anything from the lack of markets for public goods and services (called public goods failure, e.g. absence of "markets" for species conservation or for most of the regulating and supporting services of ecosystems) to imperfections in structure or process around markets which cause inefficiency and distortions (e.g. it can be argued that some price distortions in today's carbon markets are attributable to timid emissions caps).

Furthermore, there is potential for market-based instruments to produce results that are socially unacceptable – carbon markets could be said to have helped legitimize global greenhouse gas emission levels (42 billion tonnes), that are perhaps five times the Earth's ability to absorb such gases. The size of the challenge of market failure should not be underestimated: for some services (e.g. scenic beauty, hydrological functions and nutrient cycling) it is difficult even to obtain a profile of demand and supply.

There is an element of information failure here which leads to market failure. There are many cases across the globe where information failure is overcome by measures such as environmental impact assessments (EIA). They can provide arguments that lead to less destructive options being taken. The viability of road-building projects connecting Mexico and Guatemala through the Mayan forest was challenged on economic grounds. In India, information provided to the Indian Supreme Court on the value of ecosystems and biodiversity helped enshrine rates of compensation for forest conversion that will make it more difficult for approving authorities to take decisions that destroy public value.

Nevertheless, information failure is common. For example, local authorities grant land conversion permits that lead to the fragmentation of habitats or damage to ecosystems for marginal private economic gain. Decision makers often have insufficient facts, tools, arguments or support to take a different decision and avoid biodiversity loss. This is particularly unfortunate since much of the lost bio-diversity was of greater benefit to the region than the private gains.

There are many cases of local economy and local societal losses in the interests of short-term private gain. Lack of secure property rights is another cause of market failure. Many people in developing countries may have weak legal rights over the lands on which they live and work. This may become an incentive to "mine" these lands rather than to manage them sustainably.

**Failures of Policy**

Policy failures arise due to incentives encouraging harmful action. Tax incentives and subsidies can lead to the market working for the destruction of natural capital, even where natural assets offer a sustainable flow of services to the economy and to society. Environmentally harmful subsidies discriminate against sound environmental practices while encouraging other, less desirable activities. Fisheries are an example of this.

Such subsidies are often economically inefficient, prompting growing calls for reform. Policy failures also arise when the system of incentives fails to reward those who work to improve the environment, or fails to penalize those who damage it. Many agricultural practices can support high-value biodiversity. But without appropriate recognition, for example through payments for environmental services (PES), some good practices risk disappearing.

There are often no mechanisms for winning compensation from those who damage the environment for those who have lost as a result. Upstream mining activities do not generally pay those downstream for the fish they can no longer eat, or for health impacts. While such failures are still the norm, there is a shift in some countries. Costa Rica is the poster child for PES, although the approach is widely used in developed countries as well in the form of agri-environment subsidies.

Overall, benefit sharing is becoming a more acceptable concept, and liability and compensation payments are sometimes offered at levels that begin to act as real incentives. We elaborate on these aspects in the following chapter. Lastly, due to population pressures, poverty and weak enforcement of protection, development policies sometimes indirectly result in natural ecosystems being converted into agricultural or urban landscapes in situations where, for social and environmental reasons, these are not the optimal choices.

This is an example of policy failure driven by institutional failure and information failure. Formal and informal networks and rules are needed to support responses to policies which effectively manage ecosystem services. The costs of such institutional frameworks can be called policy costs and we return to this topic later in the chapter. But before we discuss and analyse benefits and costs, we would like to recognize three important issues – risks, uncertainty, and the principle of equity – which must be addressed.

Not only do they influence analysis, evaluation and the design of solutions for the various failures we have outlined above, but because they are in essence deep ethical issues, they translate into underlying assumptions for our analytical framework. We show that selecting an appropriate discount rate, a vital component of any cost benefit analysis, is the outcome of implicit or explicit ethical choices.

## ECONOMICS, ETHICS AND EQUITY

Economics has developed techniques to deal with risks, uncertainty and questions of equity. Discounting is a key tool in many conventional economic analyses because it helps to assess the value of cash flows resulting from decisions taken now. Conventional economic approaches can also be important in valuating biodiversity, but they cannot necessarily be applied routinely because of the potentially extreme consequences of biodiversity decisions. We outline below the complexities of applying economics in a field such as biodiversity.

### The Effect of Subsidies on Fisheries

Subsidies are considered to be one of the most significant drivers of over fishing and thus indirect drivers of degradation and depletion in marine biodiversity.

- Subsidies fund fisheries expansion. Globally, the provision of subsidies to the fisheries industry has been estimated at up to US$ 20-50 billion annually, the latter roughly equivalent to the landed value of the catch.
- Over half the subsidies in the North Atlantic have negative effects through fleet development. This includes decommissioning subsidies, which have been shown usually to have the effect of modernizing fleets, thereby bringing about an increase in their catching powers.
- While fishing vessel populations stabilized in the late 1990s, cheap fuel subsidies keep fleets operating even when fish are scarce.
- The Common Fisheries Policy of the European Community, for example, allows for vessels to be decommissioned to reduce effort in some countries while simultaneously subsidizing others to increase their fishing capacity.

## RECOGNIZING RISKS AND UNCERTAINTY

The treatment of climate change by the *Stern Review* surfaced an issue which had been widely recognized but not tackled squarely: how to assess a roll of the dice, when one of the outcomes is the end of civilization as we know it? This dilemma also applies to assessing the risks of Ecosystem collapse. The difficulty was highlighted when one academic study estimated the economic value of ecosystem services at US$ 33 trillion.

This result was criticized on the one hand for being far too high, but on the other hand for being "a significant underestimate of infinity". Expressed in the language of finance, the global economy is "short an option" on climate change and on biodiversity and needs to pay a premium to buy protection. The *Stern Review*'s most quoted result, that a 1% per annum cost would be needed to protect the world economy from a loss of up to 20% of global consumption, is an example of such an "option premium".

In the case of biodiversity and ecosystem losses, the size of such premiums will depend on several aspects of the ecosystem in question: its current state, the threshold state at which it fails to deliver ecosystem services, its targeted conservation state, and our best estimate of uncertainties. This is an exceedingly complex exercise as there are no market values for any of these measures.

The alarming risks of "business as usual": the loss of freshwater due to deforestation, soil erosion and nutrient loss, losses in farm productivity, the loss of fisheries; health problems and poverty. Attempting to value these losses raises important ethical dimensions – especially about the value of human well-being in the future compared to now. We believe the economics of uncertainty and discounting can help to address these ethical issues.

## DISCOUNT RATES AND ETHICS

We are addressing issues here (such as species extinction) where there is no universal agreement on the appropriate ethics. But the ethical nature of the issue is widely recognized. A group of ethics experts recently framed the issue like this: "If human behaviour is the root cause of the biodiversity extinction crisis, it follows that ethics – the inquiry into what people and societies consider to be the right thing to do in a given situation – must be part of the solution.

However, ethics is rarely accepted as an essential ingredient and is usually dismissed as being too theoretical a matter to help with the urgent and practical problems confronting conservationists." Economists discount any future benefit when comparing it to a current benefit. At one level, this is just a mathematical expression of the commonsense view that a benefit today is worth more than the same benefit in the future. But ethical considerations arise, for example when we consider giving up current income for the benefit of future generations, or the opposite: gaining benefits now at the expense of future generations.

Financial discount rates consider only the time value of money, or the price for its scarcity, and relate the present value of a future cash flow to its nominal or future value. Simple discount rates for goods and services consider just time preference, or the preference for a benefit today versus later. Social discount rates are more complex, and engage ethical aspects of a difficult choice: consumption now versus later, for society rather than for an individual.

The preferences built into this choice cover the relative value of goods or services in the future when their benefit may be lower, or higher, than now, and that benefit might flow to a different person or to a future generation.

## DISCOUNTING AND INTERGENERATIONAL EQUITY

The *Stern Review* has highlighted the crucial importance of the choice of discount rates in longterm decisions that range beyond conventional economic calculations. The discount rate has even been described as the "biggest uncertainty of all in the economics of climate change".

### Discounting and the Optimist's Paradox

There are two main reasons for discounting. The first is called "pure time preference" by economists. It refers to the inclination of individuals to prefer 100 units of purchasing power today to 101, or 105, or even 110 next year, not because of price inflation (which is excluded from the reasoning) but because of the risk of becoming ill or dying and not being able to enjoy next year's income. Whatever the reason for this attitude, it should not apply to a nation or to human society with a time horizon in the thousands or hundreds of thousands of years. Economists have often criticized "pure time preference".

The most famous critique against it was perhaps that of the Cambridge economist Frank Ramsey, in 1928. In the context of growth theory, economists agree with the discounting of the future for other reasons. They might agree with Ramsey, that to discount later enjoyments in comparison with earlier ones is "a practice which is ethically indefensible and arises merely from the weakness of the imagination".

But discount they will, as Ramsey himself did, because they assume that today's investments and technical change will produce economic growth. Our descendants will be richer than we are. They will have three, four or even more cars per family. Therefore, the marginal utility, or incremental satisfaction they will get from the third, fourth or fifth car, will be lower and lower. Discounting at the rate at which marginal utility decreases could be ethically justified.

Growth is then the reason for undervaluing future consumption and future enjoyments. Is it also a reason to undervalue future needs for environmental goods and services? It is not, particularly if we think of irreversible events. Economic growth might produce virtual Jurassic Theme Parks for children and adults; it will never resurrect the tiger if and when it goes. Growth theory is economic theory. It does not take out from the accounts the loss of nature, nor does it exclude from the accounts the defensive expenditures by which we try to compensate for nature's loss (building dykes against sea-level rise induced by climate change, or selling bottled water in polluted areas).

If we try to add up the genuine increase of the economy because of positive technical changes and investments (which nobody would deny), and the loss of environmental services caused by economic growth, the balance would be doubtful. In fact, we step on the issue of incommensurability of values. Discounting gives rise to "the optimist's paradox".

Modern economists favour discounting not because of "pure time preference" but because of the decreasing marginal utility of consumption as growth takes place. The assumption of growth (measured by GDP) justifies our using more resources and polluting more now than we would otherwise do. Therefore our descendants, who by assumption are supposed to be better off than ourselves, perhaps will be paradoxically worse off from the environmental point of view than we are.

This is because the events being considered will happen over periods of 50 years or more, and the effect of choosing different discount rates over such long periods is significant. The effects of only small differences in the discount rate, applied to a cash flow of US$ 1 million in 50 years' time, are dramatic.

A zero discount ratemeans the cost or benefit is worth the same now as it would be in 50 years, but small increases in the rate result in substantial reductions in the present value of the future cash flow. An annual discount rate of 0.1% produces a present value of 95% of the forward cash flow (US$ 951,253). Discounted at 4%, the result is only 14% of the future cash flow – just US$ 140,713.

**Table. Discount Rates and Outcomes**

| 50-year forward cash flow | Annual discount rate% | Present value of future cash flow |
|---|---|---|
| 1,000,000 | 4 | 140,713 |
| 1,000,000 | 2 | 371,528 |
| 1,000,000 | 1 | 608,039 |
| 1,000,000 | 0.1 | 951,253 |
| 1,000,000 | 0 | 1,000,000 |

Applying a 4% discount rate over 50 years implies that we value a future biodiversity or ecosystem benefit to our grandchildren at only one-seventh of the current value that we derive from it! If our ethical approach sees our grandchildren valuing nature similarly to our generation, and deserving as much as we do, the discount rate for valuing such benefits over such a time period should be zero. Unlike man-made goods and services which are growing in quantity (hence the argument to discount future units of the same utility), the services of nature are not in fact likely to be produced in larger quantities in future.

Perhaps the discount rate for biodiversity and ecosystem benefits should even be negative, on the basis that future generations will be poorer in environmental terms than those living today, as Paul Ehrlich has suggested. That raises important questions about present policies which assume significant positive discount rates.

When incomes are expected to grow, goods or services delivered later are relatively less valuable (because they represent a smaller part of the future income). This supports the usual, positive discount factor. The opposite holds true when asset values or incomes are expected to fall – future goods and services will become more valuable than now. In the case of biodiversity it is questionable whether it will be equally, more or less available in future, and therefore even the direction of the discount rate is uncertain.

## DISCOUNTING IN A WELFARE CONTEXT

In welfare economics the objective is to maximize the social benefits of consumption across all individuals, with "consumption" covering a broad range of goods and services, including health, education and the environment. Aggregating social utility across individuals is problematic and prone to value judgements such as comparing the value of consumption for a rich person versus a poor person. What are "appropriate" discount rates for communities or countries with significant poverty and hardship?

Focusing on poverty alleviation now means that the benefits and costs of today's poor are more valuable than those of future generations (who may live under better conditions). This is an ethical argument for high discount rates! But if today's poor rely directly on the conservation of biodiversity for vital supplies such as freshwater and fuel wood, is it then justifiable to provide more income options to today's rich if this would jeopardize these vital supplies? Consider some examples of ethically indefensible trade-offs.

A forest ecosystem may be essential to the well-being of poor farming communities downstream – by providing nutrient flows, recharging aquifers, regulating seasonal water supply, preventing soil erosion and containing flood damage and drought losses. It could be ethically difficult to justify destroying such a forest watershed in order to release economic value which has utility for the agents of destruction (e.g. profits from minerals and timber, related employment, etc.), whilst on the other hand, the costs of replacing ecosystem benefits forgone may be the same or less in monetary terms, but impossible to bear in human terms as they fall on poor subsistence farming communities. We see such situations as outcomes of bad economic targeting – *economics is mere weaponry, its targets are ethical choices.*

## "GDP of the Poor"

The full economic significance of biodiversity and ecosystems does not figure in GDP statistics, but indirectly its contribution to livelihood and well-being can be estimated and recognized. Conversely, the real costs of depletion or degradation of natural capital (water availability, water quality, forest biomass, soil fertility, topsoil, inclement microclimates, etc.) are felt at the micro-level but are not recorded or brought to the attention of policy makers. If one accounts for the agricultural, animal husbandry and forestry sectors properly, the significant losses of natural capital observed have huge impacts on the productivity and risks in these sectors.

Collectively, we call these sectors (i.e. agriculture, animal husbandry, informal forestry) the "GDP of the poor" because it is from these sectors that many of the developing world's poor draw their livelihood and employment. Furthermore, we find that the impact of ecosystem degradation and biodiversity loss affects that proportion of GDP most which we term the "GDP of the poor". The end-use of ecosystem and biodiversity valuations in National Income Accounting, either through satellite accounts (physical and monetary) or in adjusted GDP accounts ("Green Accounts") does not of itself ensure that policymakers read the right signals for significant policy trade-offs.

A "beneficiary focus" helps better recognize the human significance of these losses. In exploring an example for this interim report, we found that the most significant beneficiaries of forest biodiversity and ecosystem services are the poor, and the predominant economic impact of a loss or denial of these inputs is to the income security and well-being of the poor. An "equity" focus accentuated this finding even further, because the poverty of the beneficiaries makes these ecosystem service losses even more acute as a proportion of their livelihood incomes than is the case for the people of India at large.

We find that the per-capita "GDP of the poor" for India increases from US$ 60 to US$ 95 after accounting for the value of ecological services, and also that if these services were denied, the cost of replacing lost livelihood, equity adjusted, would be US$ 120 per capita – further evidence of the "vicious cycle" of poverty and environmental degradation.

We shall explore this approach for the developing world more broadly in Phase II. We believe that by using such sectoral measures and forcing a reflection of the equity principle by its "human" significance (given that most of the world's 70% poor are dependent on this sector) we shall focus adequate importance on policy making and contribute to a halt in the loss of biodiversity.

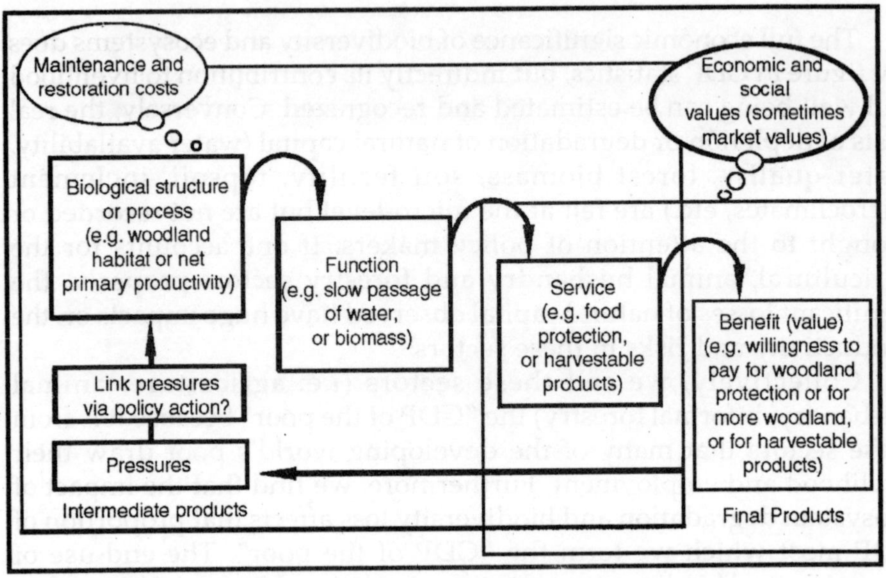


**Fig.** The link Between Biodiversity and the Output of Ecosystem Services

## DISCOUNTING BIODIVERSITY LOSSES

We do not suggest that there are always defensible "tradeoffs" for ecosystems and biodiversity, especially if significant ecosystems cease to function altogether as providers of provisioning or regulating services, or if biodiversity suffers significant extinctions. The evaluation of trade-offs using cost-benefit analysis and discounting works best for marginal choices involving small perturbations about a common growth path. However, the reality is that there are trade-offs, explicit or implicit, in any human choice.

Even trying to set a boundary where trade-offs should not apply is itself a trade-off! Trade-offs involve a choice between alternatives, and in the case of biodiversity losses, there are not always comparable alternatives. For development to be considered sustainable, a boundary condition called "weak sustainability" is defined, being a situation in which overall capital – natural, human and physical – is not diminished. But this also suggests that one form of capital can be substituted for another, which is not true: more physical wealth cannot always be a substitute for a healthy environment, nor vice versa.

However, it is important for all aspects of the "natural capital" side of a trade-off at least to be appropriately recognized, valued and reflected in cost-benefit analysis, and even this is not yet being done in most trade-off decisions.

There is a different boundary condition called "strong sustainability" which requires no net diminution of natural capital: this is more difficult to achieve, although compensatory afforestation schemes are examples of instruments designed to achieve strong sustainability.

Finally, any trade-off has to be ethically defensible, and not just economically sound. With biodiversity, we are not only considering long-term horizons as we are with climate change.

Ecosystem degradation is already extensive and observable, and some of its effects are dramatic – such as the loss of freshwater causing international tension.

Significant biodiversity losses and extinctions are happening right now, and flagship species such as the Royal Bengal tiger in India are under threat. A higher or lower discount rate can change the quantification of the social cost of imminent losses, but it would not alter the nature of the outcomes – loss of vital ecosystem services and valuable biodiversity.

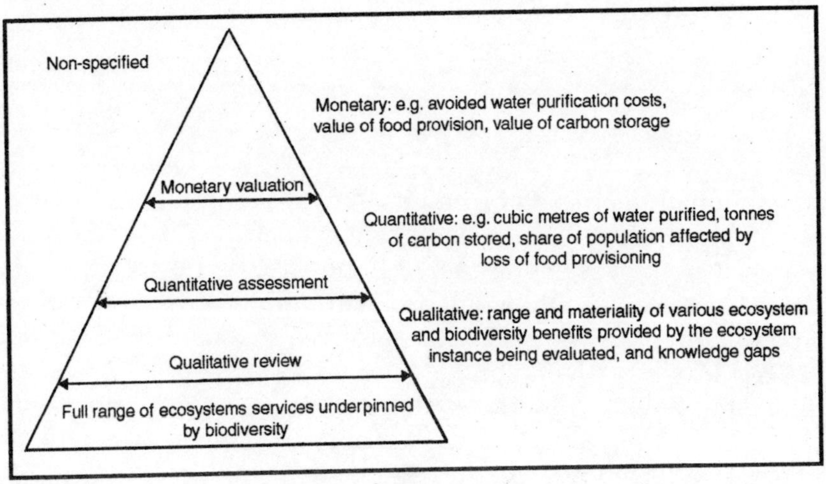


**Fig.** Valuing Ecosystem Services

In one of the accompanying papers of Phase I, approximately 200 valuation studies on forests have been examined. Many of these included some discounting of annuity flows in order to calculate an aggregate value for natural capital.

We found that most studies used social discount rates of 3-5% or higher, and that none were below 3%. Our intention in Phase II is to leverage off this body of work, but to recalculate its results with different discounting assumptions.

Thus in Phase II we will propose a conceptual framework for the economics of biodiversity and ecosystem valuation which includes assessments of the sensitivity of ecosystem values to ethical choices. Our intention is to present a discrete range of discounting choices connected to different ethical standpoints, enabling endusers to make a conscious choice.

## Chapter 7

# The Evaluation Challenge

Economic evaluation can shed light on trade-offs by comparing benefits and costs and taking account of risks, and this can be applied to alternative uses of ecosystems. But there are many difficulties, which we set out in this section, and which we will address in Phase II.

Before economic valuation can be applied it is necessary to assess ecosystem changes in biophysical terms. Most benefits provided by ecosystems are indirect and result from complex ecological processes that often involve long lag times as well as non-linear changes.

Pressures may build up gradually until a certain threshold is reached, leading to the collapse of certain functions. A typical example is forest die-back caused by acidification.

The impacts of pressures on ecosystems, including the role of individual species, the importance of overall levels of biodiversity, the relationship between the physical and the biological components of the ecosystem, and the consequences with regard to the provision of services, are difficult to predict.

Economic valuation builds on the biophysical understanding and aims to measure people's preferences for the benefits from ecosystem processes. These benefits may accrue to different categories of population over different geographical and time scales.

Our ability to assess the benefits provided by ecosystems, or the costs from their loss, is limited by lack of information at several levels. There are probably benefits that we have not yet identified; so we are able to assess, even in qualitative terms, only part of the full range of ecosystem services. We will probably never be able to assess the full range.

It will be possible to make a quantitative assessment in biophysical terms only for part of these services – those for which the ecological "production functions" are relatively well understood and for which sufficient data are available. Due to the limitation of our economic tools, a still smaller share of these services can be valued in monetary terms.

## Putting it Together – an Example of a Cost of Policy Inaction Study on Biodiversity Loss

In November 2007, a consortium1 started work on a "Cost of Policy Inaction" or COPI study on the costs of not halting biodiversity loss. COPI's approach is the mirror image of benefits valuation, with the use of scenario analysis. Their terms of reference were to build up a global quantitative picture between now and 2050, and to try to value this in monetary terms. The project succeeded in establishing an appropriate approach, identifying the data gaps and the methodological problems, and providing indicative figures. Some interesting, be they only illustrative, results have been produced.

## MODELLING BIODIVERSITY LOSS

The GLOBIO model was used to project changes in terrestrial biodiversity to 2050. The main indicators were changes in land use and quality and the mean abundance of the original species of an ecosystem (MSA), for all of the world's biomes. The model provides regional estimates for conversions from natural to managed forest and from extensive to intensive agriculture, and for the resulting decline in natural areas.

The largest driver for conversions has historically been demand for agricultural land and timber, although infrastructure development, fragmentation, and climate change are predicted to become increasingly important. The expected loss of biodiversity by 2050 is about 10-15%(decline inMSA), the most extreme being in savannah and grassland.

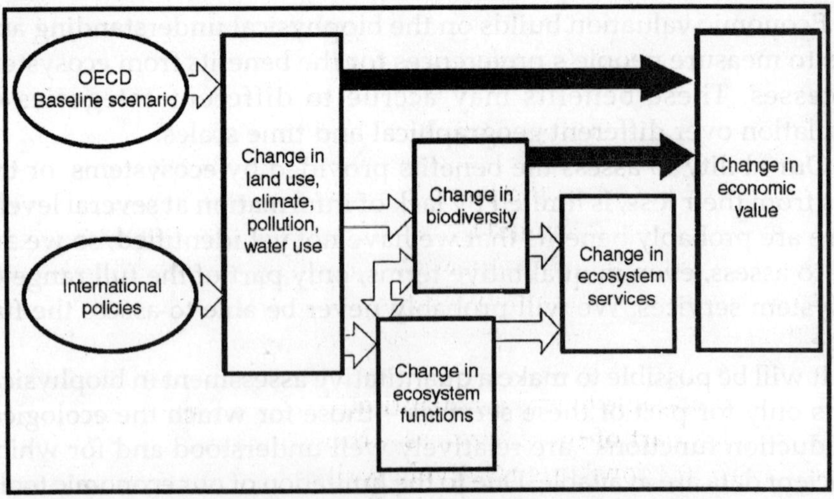


Fig. Establishing a Scenario Analysis

It is therefore important not to limit assessments to monetary values, but to include qualitative analysis and physical indicators as well. The "pyramid" diagram in Figure illustrates this important point. Measurement approaches vary depending on what we measure. For provisioning services (fuel, fibre, food, medicinal plants, etc.), measuring economic values is relatively straightforward, as these services are largely traded on markets.

The market prices of commodities such as timber, agricultural crops or fish provide a tangible basis for economic valuation, even though they may be significantly distorted by externalities or government interventions and may require some adjustments when making international comparisons. For regulating and cultural services, which generally do not have any market price (with exceptions such as carbon sequestration) economic valuation is more difficult.

However, a set of techniques has been used for decades to estimate non-market values of environmental goods, based either on some market information that is indirectly related to the service (revealed preference methods) or on simulated markets (stated preference methods). These techniques have been applied convincingly to many components of biodiversity and ecosystem services.

But they remain controversial. Fundamentally, there is the ethical question about the extent to which some life-supporting functions of biodiversity can be fully addressed by economic valuation and be considered as part of possible trade-offs instead of being dealt with as ecological constraints. Similarly, economic valuation may not be appropriate to address spiritual values.

Keeping these limitations in mind, substantial progress has been made since the 1990s by economists working with natural scientists to improve these methods: there is increasing consensus on the conditions under which they can be used, and increasing confidence in the comparability of the results. These techniques are now commonly applied to measure a wide range of values, including many indirect and non-use values. Another set of challenges relates to assessing the consequences of the loss of biodiversity and ecosystem services on a large scale. First, valuation methods generally do not cover second-round effects of the losses on the wider economy.

To assess such effects, the use of economic models is necessary. While there are already some promising attempts, this is still very much an area of ongoing research. Secondly, most of the valuation evidence comes from individual case studies concerning a particular ecosystem or species. Some studies have tried to make a global assessment of the world's ecosystem services but, while they have been useful in raising attention and discussion, their results are controversial. Others focus at species or genera levels.

Any integral assessment on a broad scale raises substantial difficulties: how to define a coherent framework; how to deal with limitations in data; how to aggregate values to estimate the global impacts of large scale changes in ecosystems. In Phase II, we expect to rely on "benefit transfer" logic, i.e. using a value estimated in a particular site as an approximation of the value of the same ecosystem services in another site. Benefit transfer is easier for some homogeneous values (such as carbon absorption, which is a global good), than for others that are site-specific or context-dependent (such as watershed protection).

However, we must recognize the trade-off between providing incomplete assessment on the one hand, and using inferred estimates (rather than primary research-based estimates) on the other. For both ecological and economic reasons, caution is needed when scaling up and aggregating values estimated from small marginal changes to assess the effects of large changes. Ecosystems often respond to stress in a non-linear fashion.

Large changes in ecosystem size or condition may have abrupt effects on their functioning, which may not be extrapolated easily from the effect of small changes. Generally, as some ecosystem services decline substantially as we continue to use them, extrapolation of benefits should recognize and be limited by the "law of diminishing returns".

**The multiple Values of Coral Reefs**

Coral reefs provide a wide range of services to around 500 million people. Some 9-12% of the world's fisheries are based directly on reefs, while a large number of offshore fisheries also rely on them as breeding, nursery or feeding grounds. Tourism generally is the dominant benefit. Reef recreation has been estimated at US$ 184 per visit globally, at US$ 231-2,700 per hectare per year in Southeast Asia and at US$ 1,654 per hectare per year in the Caribbean.

Coral reefs provide genetic resources for medical research, and ornamental fish and pearl culture are extremely important for the economies of some insular states, such as French Polynesia. The reefs protect coastal areas in many islands: this vital service has been estimated to be worth US$ 55-1,100 per hectare per year in Southeast Asia. The scenario used was largely developed by the OECD as its baseline.

It is broadly consistent with other modelling exercises such as those by the FAO or other UN agencies. The model itself forecasts a slowing rate of biodiversity loss in Europe (compared to an increasing rate worldwide).

### Assessing Changes in Ecosystem Services and Applying Monetary Values

Changes in land use and biodiversity are translated into changes in ecosystem services. The assessment relies to a large extent on the valuation literature, and creative solutions have been developed to extrapolate and fill data gaps. This is an area where further work is clearly needed in Phase II. The biggest difficulty has been to find studies to monetize changes in ecosystem services.

While there are many case studies, not all regions, ecosystems and services are equally covered, and there were often difficulties in identifying values per hectare for use in such a wide spread benefit transfer. Also, most studies are based on marginal losses, and the values are often location specific.

## THE VALUATION RESULTS

In the first years of the period 2000 to 2050, it is estimated that each year we are losing ecosystem services with a value equivalent to around EUR 50 billion from land-based ecosystems alone (it has to be noted that this is a welfare loss, not a GDP loss, as a large part of these benefits is currently not included in GDP).

Losses of our natural capital stock are felt not only in the year of the loss, but continue over time, and are added to by losses in subsequent years of more biodiversity. These cumulative welfare losses could be equivalent to 7% of annual consumption by 2050. This is a conservative estimate, because:

- It is partial, excluding numerous known loss categories, for example all marine biodiversity, deserts, the Arctic and Antarctic; some ecosystem services are excluded as well (disease regulation, pollination, ornamental services, etc.), while others are barely represented (e.g. erosion control), or under-represented (e.g. tourism); losses from invasive alien species are also excluded.
- Estimates for the rate of land-use change and biodiversity loss are globally quite conservative.
- The negative feedback effects of biodiversity and ecosystem loss on GDP growth are not fully accounted for in the model.
- Values do not account for non-linearities and threshold effects in ecosystem functioning.

## CONCLUSIONS AND NEXT STEPS

The study showed that the problem is potentially severe and economically significant, but that we know relatively little both ecologically and economically about the impacts of future biodiversity loss. Further work is envisaged in Phase II to address the points mentioned above, and further elaborate on the framework and methodology in line with our recommendations.

## THE COSTS OF BIODIVERSITY LOSS

There is a substantial body of evidence on the monetary values attached to biodiversity and ecosystems, and thus on the costs of their loss. A number of recent case studies and more general contributions have been received in reply to a call for evidence

**Table. Projection of Total Benefits of Carbon Storage in European Forests.**

| | Latitude | | | |
|---|---|---|---|---|
| | **35-45** | **45-55** | **55-65** | **65-71** |
| Value per hectare (US$, 2005) | 728.56 | 1,272.85 | 468.60 | 253.33 |

Our Phase I report COPI made a first review of the general valuation literature and databases and tried to build up a global quantitative picture of biodiversity loss in biophysical and monetary terms. A more targeted review of valuation case studies concerning forest ecosystems has also been made.

The existing valuation studies vary in their scope, quality, methodology, and suitability for use in a large-scale assessment. Often the estimated economic values are not comparable as they may be of a different nature or be expressed in different units, or the estimates may not be clearly related to a specific service or an area. A particular effort is needed to assess indirect use values, especially those of regulating services, which are receiving increasing attention as a consequence of the Millennium Ecosystem Assessment.

For carbon sequestration, substantial values have often been found, although they vary depending on the type of forest – for example deciduous or coniferous – and their geographical location. Some significant values have been estimated for water regulation, although they are highly context-specific. The value of the watershed protection provided by intact coastal ecosystems, such as mangroves and other wetlands, has been estimated at US$ 845 per hectare per year in

Malaysia and US$ 1,022 per hectare per year in Hawaii, United States of America.

Overall, the values of the multiple watershed services tend to range from US$ 200 to 1,000 per hectare per year. The value of bee pollination for coffee production has been estimated at US$ 361 per hectare per year, although the benefits only accrued to producers within 1 kilometre of natural forests. Many of the studies evaluating regulating services, for example for coastal protection or regulation of the water cycle, use production function approaches.

These approaches are being increasingly refined, allowing better assessment of trade-offs between competing uses of ecosystems. While there is increasing evidence of the value of some regulating services, many others, such as health regulation, have been little explored so far, although there are some indications that they might be significant. The economic importance of the contribution of aggregated biodiversity to ecosystem resilience (the capacity of an ecosystem to absorb shocks and stresses in constructive ways) is probably very high but still poorly quantified, although studies have analysed aspects such as the contribution of crop diversity to agricultural yields and farm income.

This important gap in knowledge reflects the difficulty of first quantifying the risks of a system collapse from an ecological perspective, and then measuring people's willingness to pay to reduce those risks which are not yet well understood. The real costs of the loss of biodiversity and ecosystems also include option values.

Although they are difficult to measure, these values placed on conserving resources for possible uses in the future are significant because our knowledge of the importance of ecosystem services is expected to improve over time, and because part of the losses of biodiversity and the services it underpins are irreversible. A preferred methodology for measuring option values (in particular, bio-prospecting values) has been prepared as part of the preparatory work in Phase I. In Phase II, we propose to build on this approach.

## THE COSTS OF BIODIVERSITY CONSERVATION

Losing biodiversity and ecosystem services might cause tremendous costs for society due to the subsequent loss of various provisional and regulatory services such as food production, water regulation and climate change resilience. All these create the necessary arguments for biodiversity protection, while the rate of loss demands urgent action. But conservation, too, has a cost, which needs to be factored into decision making. Knowing these costs provides the basis for determining the relationship between costs and benefits, and for identifying the most cost-effective options of conservation.

**Table. Results from Studies on the Costs of Conservation**

| Object | Assessed costs | Estimates |
|---|---|---|
| Conserving the Cape Floristic Region (South Africa) | OC + MC | One-off US$ 522 million and annual expenses of US$ 24.4 million |
| Network of protected ecosystems (Bahia, Brazil) | OC | OC 10.000 ha |
| Preservation of tropical forest (certain regions) | OC | Sumatra: US$ 0.95/ha/year<br>Borneo: US$ 1.10/ha/year<br>Sulawesi: US$ 0.76/ha/year<br>Java/Bali: US$ 7.82/ha/year<br>Malaysia: US$ 27.46/ha/year |
| Non-timber forest product benefits (Nagarhole National Park, India) | OC | Net present value of US$ 28.23 per household annually |
| Biodiversity protection (Brigalow Belt, New South Wales) | OC | US$ 148.5 million |
| Biodiversity protection within the Natura 2000 network (covering 18% of EU25-territory) | MC + TC | EUR 6.1 billion annually over a 10-year period |
| Expanding forest conservation to all priority areas (worldwide) | OC + MC | US$ 5.75/ha/year for 10 years |

A comprehensive cost assessment has to include various types of costs: biodiversity conservation may require use restrictions which lead to opportunity costs from foregone economic development; management costs arise for measures like, for example, fencing and breeding programmes, and transaction costs are associated with the design, implementation, and control policies for biodiversity conservation.

Globally, between US$ 8 billion and US$ 10 billion are invested annually in biodiversity conservation; protected areas take up a significant portion of these resources. At the global level, US$ 28 billion may be required annually over the next 30 years to expand IUCN priority habitats to 10% of the area of all countries. This cost estimate includes acquisition and management costs of the current and future biodiversity reserve sites. If the protected area system is expanded to cover key currently unprotected species and to meet biological/ ecological needs, up to US$ 22 billion per year in management costs would be required.

But safeguarding the provisioning of ecosystem services and biodiversity benefits in protected areas could cost as little as two orders of magnitude less than the valued benefits of ecosystems and biodiversity. Balmford et al. have taken this idea and proposed that, for the annual investment of US$ 45 billion – around a sixth of that needed to conserve all ecosystem services worldwide – we could protect natural services worth some US$ 5 trillion in protected areas: an extremely good benefit-cost ratio of 100:1.

The costs of conservation vary between regions because of differences in their economies and cost structures. Conservation costs have been found to be as little as US$ 0.01 per hectare per year in remote areas and as high as US$ 1,000 per hectare per year in densely populated areas. The benefits of services derived from different ecosystems go from several hundred to over US$ 5,000 per hectare annually and in some cases much more.

An extreme case is that of coral reefs, for which UNEP estimated an overall value of ecosystem services at between US$ 100,000 and US$ 600,000 per square kilometre; based on an estimated cost of US$ 775 per square kilometre for maintaining marine protected areas, the management costs of coral reefs could be as little as 0.2% of the value of the ecosystem protected – the opportunity costs of coral reef conservation are not included in this comparison.

However, knowledge of the spatial distribution of benefits and costs of biodiversity protection is necessary to enable cost effective conservation of ecosystem services. Although the figures available so far apply to small bits of nature here and there, policy makers want the big picture. When the Natura 2000 network of protected areas started to emerge in the European Union, one common thread was the costs of managing it and delivering on the targets. The cost of implementing this network of protected sites, which then accounted for 18% of the EU-25 territory, has been estimated at over EUR 6 billion annually.

These costs included management, restoration and provision of services (like recreation and education), but excluded expenditures for buying up land for nature. The overall costs of conservation are higher if we include philanthropy and subsidies. For example, in the United States of America, private charity to "environment and animals" was estimated at US$ 9 billion in 2005. Protected areas in developing countries are considerably cheaper per hectare to establish and manage than in developed countries.

Thus, although developing countries account for 60%of the total area of biodiversity reserve sites, their actual conservation budget needs come to just 10%of the global budget. The costs of achieving a given conservation target depend on the chosen policy instruments and their designs. In testing this assumption, it was found that simply using a different design for a conservation instrument can yield cost-savings of up to 80% for a given species coverage. A necessary, but not sufficient, prerequisite for cost-effective spending is that conservation spending conforms to current conservation priorities. Only 2-32% of spending patterns by conservation agencies can be explained by the guidelines for prioritizing biodiversity conservation.

Another point to be considered is the spread of resources needed between different portions of biodiversity. In economic terms, it appears that the marginal costs of conservation investments are increasing: that is, while the first "units" of conservation can be bought at a low cost, each additional unit costs more. However, researchers believe that "low-hanging fruits" are available in biodiversity conservation. Saving a large number of species is relatively cheap, but costs may explode as the last few species, habitats or ecosystems are included in conservation targets.

The general scarcity of studies pointing to the benefits and costs of biodiversity conservation, especially at regional and local levels, contributes to the non-allocation of sufficient resources for conservation and to the observed budget shortfalls. Only a very limited number of studies have assessed simultaneously the benefits and costs of protecting biodiversity and ecosystem services in specific conservation projects.

Some studies have been area-specific, like the assessment of protecting ecosystem services in Madagascar, which revealed that the country's biodiversity provides a wide array of services that brings benefits of a value twice as high as the management costs of biodiversity resources on the island. Other studies have been sector-specific: for example, it has been estimated that a global marine protected area system, accounting for the closure of 20%of total fishing area and resulting in a lost profit of US$ 270 million per year, would help sustain fisheries worth US$ 70-80 billion per year while creating 1million jobs.

Furthermore, the methodology used for studies on conservation costs often lacks some common understanding of what to include and how to measure these costs. The resulting picture of conservation economics is incomplete and a spatially explicit method for distributing conservation funding is missing. Although conservation of biodiversity appears economically reasonable, current global expenditures (estimated at US$ 10-12 billion annually) fall short of expected needs.

Because conservation, mainly in developing countries, suffers from budget shortfalls, developing countries should get priority when allocating additional money for global biodiversity conservation to enhance the effectiveness of their protection measures.

However, as conservation goals for developing countries are often seen to compete with their development goals, there are important societal issues to address in a local context: property rights versus access rights and usufruct rights, rights for local residents versus rights for migrant and neighbouring poor, livelihood and welfare issues, and the persistence of the "vicious cycle" of poverty and environmental degradation. In addressing these issues in Phase II, we have to recognize policy overlaps that will affect the viability of an economic toolkit for policy makers in the developing world.

## PROPOSED VALUATION FRAMEWORK

The considerations explained in this chapter have led to a valuation framework which we propose to use in Phase II, in conjunction with our meta-analysis of valuation studies, so that we can prepare a globally comprehensive and spatially specific framework and estimation grid for the economic valuation of ecosystems and biodiversity. It is based on the science study1 and the issues of ethics, equity and the discount rate discussed earlier. These are the key elements of our proposed framework:

- Examine the causes of biodiversity loss: designing appropriate scenarios to evaluate the consequences of biodiversity loss means incorporating information on the drivers of this loss. For example, loss of marine fisheries is driven by over fishing, so it would be appropriate to compare a scenario of business-as usual (continued over fishing) with one where fisheries are sustainably managed. Evidence suggests that biodiversity is often being lost even where it would be socially more advantageous to preserve it. Identifying the market, information and policy failures can help us identify policy solutions.

- Evaluate alternative policies and strategies that decision makers are confronted with: the analysis needs to contrast two or more "states" or scenarios that correspond to alternative action (or inaction) to reduce biodiversity and ecosystem loss (World A and World B). This approach is also used in impact assessments and cost-benefit analyses to ensure that decision makers can make informed decisions on the basis of a systematic analysis of all the implications of various policy choices.
- Assess the costs and benefits of actions to conserve biodiversity: the analysis will need to address both differences in benefits obtained from biodiversity conservation (e.g. water purification obtained by protecting forests) and in the costs incurred (e.g. foregone benefits from conversion of the forest to agriculture).
- *Identify risks and uncertainties*: There is much that we do not know about how biodiversity is valuable to us, but that does not mean that what is not known has no value – we risk losing very important, but still unrecognized, ecosystem services. The analysis needs to identify these uncertainties and assess the risks.
- *Be spatially explicit*: Economic valuation needs to be spatially explicit because both the natural productivity of ecosystems and the value of their services vary across space. Furthermore, benefits may be enjoyed in very different places from where they are produced. For example, the forests of Madagascar have produced anti-cancer drugs that save lives all over the world. Besides, the relative scarcity of a service, as well as local socio-economic factors, may substantially affect the values. Taking into account the spatial dimension also allows for better understanding of the impacts of conservation on development goals, and for the exploration of trade-offs between the benefits and costs of different options,

highlighting regions that may be cost-effective investments for conservation.

- Consider the distribution of impacts of biodiversity loss and conservation: the beneficiaries of ecosystem services are often not the same as those who incur the costs of conservation. Mismatches can lead to decisions being taken that are right for some people locally, but wrong for others and for society as a whole. Effective and equitable policies will recognize these spatial dimensions and correct them with appropriate tools, such as payments for ecosystem services.

Figures illustrate the multi-scale dimension of ecosystem services and thus the need to account for the spatial pattern in their production and use. Even large cities like London depend on a diversity of benefits produced by ecosystems and biodiversity, often at a considerable distance. This framework will be used during Phase II but it will not be possible to collect information for the elaboration of detailed maps for all types of ecosystem services and biomes.

Thus the evaluation will also largely rely on "benefits transfer", making clear the assumptions and defining carefully the conditions for extrapolating from limited data, taking into account the scale and distance-dependency of the various services. Spatial data bases will be used, highlighting where data gaps need to be filled.

## Bringing Together the Ecological and Economic Aspects in our Valuation Framework

Valuing ecosystems requires integration of ecology and economics in an interdisciplinary framework. Ecology should provide the necessary information on the generation of ecosystem services, while economics would bring the tools for estimating their values. Valuation of regulating ecosystem services and some provisioning services must be based on an understanding of the underlying biological and physical processes that lead to their provision.

For example, to be able to valuate the water regulation provided by a forest, it is first necessary to have information about the land use, the hydrology of the area and other characteristics, in order to make a biophysical assessment of the service provided. Such an understanding makes it possible to estimate economic value, but there are some challenges which need to be addressed:

- Measuring the quantity and quality of services provided by ecosystems and biodiversity in various possible states is a key challenge, and also an opportunity, to avoid the pitfalls of generalization. Valuation is best applied to alternative states or scenarios (e.g. services provided under differing land-use practices reflecting different policy scenarios). For example, the conservation of tropical forest catchments might provide net benefits to water compared to the same area as grazing land or cropland, but these benefits might not exceed the benefits of agro-forestry on the same piece of land. Estimating extant biodiversity in these different scenarios would add another layer of challenge. It would be important to scope such scenario-based assessment appropriately, to ensure that the main purpose of our valuations (estimating the costs and benefits of conservation of biodiversity) is not lost in modelling alternative land uses.

- Non-linearity in the flow of services needs special attention. For example, recent studies on coastal mangroves in Thailand have taken into account that the ecosystem service providing coastal protection does not vary in line with the area of natural mangrove.This leads to significantly different values and policy conclusions compared to previous studies, notably on the optimal mix between conservation and development. Another important aspect is the existence of threshold effects and the need to assess how close an ecosystem may be to the collapse of certain services.There are still major gaps in scientific knowledge about the role of species in ecosystems and what the key factors are for producing flows of beneficial ecosystem services and ensuring their resilience.However, for some services, there is evidence on the influence of certain biophysical indicators (habitat areas, indicators of health, species diversity, etc.).The Scoping the Science study has reviewed the state of ecological knowledge on a series of ecosystem services and assessed the available information. The findings of this study– which will be added to in Phase II – will provide a basis for the economic evaluation by means of:
  - Building appropriate scenarios for the provision of each ecosystem service.
  - Defining for at least a set of services theme thod for generating a global quantification and mapping of service provision under different scenarios, upon which to base the economic valuation.
  - Formulating reasonable assumptions to allow for the extrapolation of values estimated for certain ecosystems so as to fill data gaps.

- The links between ecosystem processes and the benefits they provide to people vary in complexity and directness. A classification system is needed, and can be developed from the system built in the context of the Millennium Ecosystem Assessment, which can still be improved to provide a good basis for economic valuation. It appears useful to distinguish between "final" services (e.g. crop provision, clean water provision) that provide benefits directly relevant for human welfare, and "intermediate" services that serve as inputs for the production of other services (e.g. pollination, water regulation). The economic value of pollination, for example, cannot be assessed separately from that of the provision of crops. An end-user perspective must be adopted: the value of intermediate services can only be measured through their contribution to the production of end-user benefits. We intend to structure the classification of services for the evaluation in Phase II around this perspective.

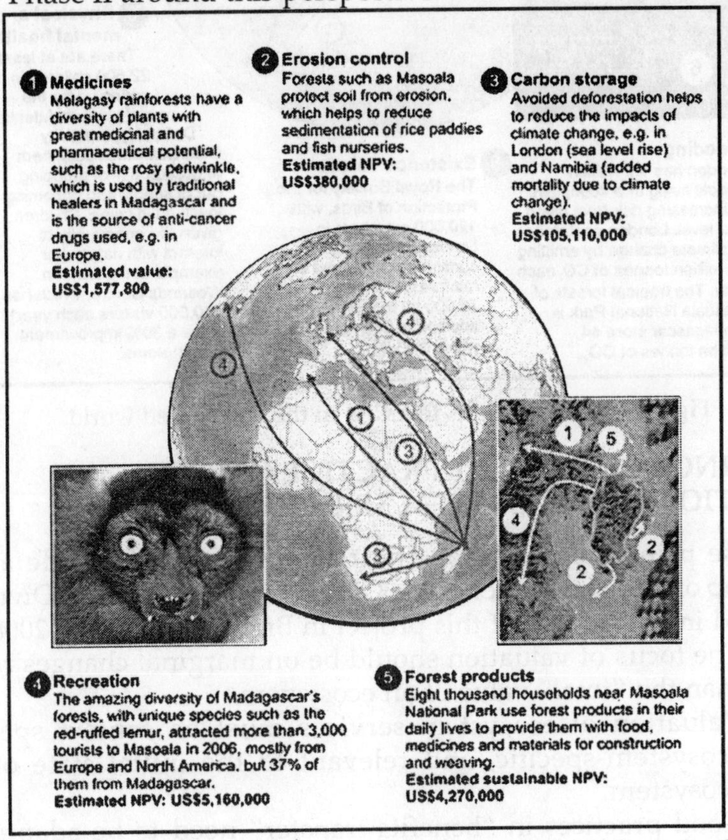


**Fig.** Ecosystem Benefits from a Protected forest in a High biodiversity County.

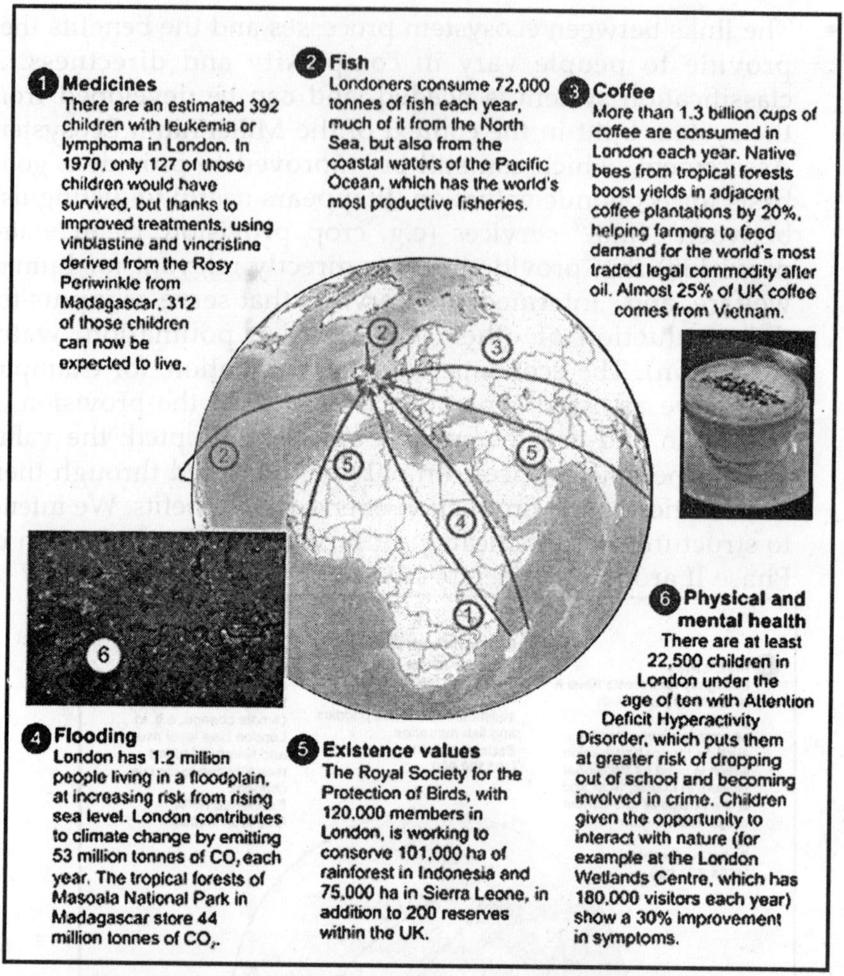


**Fig.** Ecosystem Benefits to a City in the Developed world.

## KEY PRINCIPLES OF BEST PRACTICE ON THE VALUATION OF ECOSYSTEM SERVICES

These principles build on the recommendations made at the Workshop on the Economics of the Global Loss of Biological Diversity organized in the context of this project in Brussels in March 2008.

- The focus of valuation should be on marginal changes rather than the "total" value of an ecosystem.
- Valuation of ecosystem services must be context specific, ecosystem-specific, and relevant to the initial state of the ecosystem.
- Good practices in "benefits transfer" need to be adapted to

biodiversity valuation, while more work is needed on how to aggregate the values of marginal changes.

- Values should be guided by the perception of the beneficiaries.
- Participatory approaches and ways of embedding the preferences of local communities may be used to help make valuation more accepted.
- Issues of irreversibility and resilience must be kept in mind.
- Substantiating bio-physical linkages helps the valuation exercise and contributes to its credibility.
- There are inevitable uncertainties in the valuation of ecosystem services, so a sensitivity analysis should be provided for decision makers.
- Valuation has the potential to shed light on conflicting goals and trade-offs but it should be presented in combination with other qualitative and quantitative information, and it might not be the last word.

In Phase II, we will exploit existing valuation literature in greater depth and develop a methodology for choosing valuation techniques for different benefits, and for applying benefit transfer and aggregation. The work will build on the framework described in this chapter and will refine it in these ways:

- It will focus on the contribution of services to final benefits to people, thus avoiding double counting.
- There will be a clear "spatial focus" – on the locations where the services and benefits arise.
- It will identify risks by noting the fragility of an ecosystem and whether it is judged to be near thresholds, and will reflect this in selecting a valuation approach, recognizing the limitations of conventional analysis where changes are not marginal.
- Likewise, for estimating stock values from ecosystem service flows, it will recognize the limitations of discounting where we are not looking at small variations along a given growth path.

Finally, we should affirm here that valuation is not an end in itself, and should be oriented towards the needs of endusers. This includes policy makers and decision makers at all levels of government. It also includes corporate and consumer organizations, since private-sector actors are significant users of biodiversity benefits and potential stewards of biodiversity and ecosystems.

## FROM ECONOMICS TO POLICIES

Society's defective economic compass can be repaired with appropriate economics applied to the right information. This will allow existing policies to be improved, new policies to be formed, and new markets to be created: all of which is needed to enhance human well-being and restore the planet's health.

In the last chapter we described how biodiversity is seriously affected by policy – or a lack of policy. Since there are no markets for most of the "public goods and services" from biodiversity and ecosystems, their costs and benefits often fall to different actors or at different levels, as with all "externalities".

There is little or no private reinvestment in maintaining and conserving these resources. The polluter often does not pay for causing losses to others. The subsidized fishing fleet depletes fish stocks well beyond levels that would occur in the absence of such subsidies. Vital services of forests – such as water provisioning and regulation, soil retention, nutrient flow, enhanced landscapes – do not reward the beneficiaries and are provided at much lower levels than desirable.

The benefit of conserving a species for future generations is global, whereas costs for its conservation are local and uncompensated, and therefore it becomes extinct. Despite all these "disconnects", there is room for optimism. During our Phase I studies, we observed several good policies already in place, in many countries, that address these issues. However, amore thorough consideration of the economics of biodiversity and ecosystem services is needed to make these solutions scalable and workable beyond their initial stages, "pilot" phases and current host locations.

The final report on The Economics of Ecosystems and Biodiversity (TEEB) will systematically address a comprehensive range of such policy options for better conserving biodiversity and ecosystem services, and will demonstrate how better policies result from applying and integrating the new economics of ecosystems and biodiversity. Here, we provide some examples to illustrate how the economic values of ecosystem benefits and costs can be internalized and used to help improve current policies or offer new options.

The examples come from diverse policy fields but they convey four broad messages, elaborated in the sections that follow:

- Rethink today's subsidies to reflect tomorrow's priorities.
- Reward unrecognized benefits, penalize un captured costs.
- Share the benefits of conservation.
- Measure what you manage.

## RETHINKING TODAY'S SUBSIDIES TO REFLECT TOMORROW'S PRIORITIES

Subsidies exist across the globe and across the economy. They affect us all and many affect the health of the planet's ecosystems. Harmful subsidies must be reformed to halt biodiversity loss and achieve appropriate stewardship of the planet's resources. Subsidies can support social and environmental innovation as well as technological and economic development. On the other hand they can result in private gain with no social benefits and can lead to economic inefficiency and market distortions.

Worse, they can result in biodiversity losses and damage to ecosystems. In some cases, rational support for a social objective such as food security outlives its purpose, resulting in unnecessary economic and environmental costs. Most subsidies are intentional and introduced for a clear, specific purpose, such as payments to develop commercial nuclear power in the 1950s and 1960s, and agricultural support to rebuild devastated European agriculture after World War II.

Many are permanent features – agricultural inputs and products are often subsidized directly, along with energy, food, transport and water.

### Environmentally Harmful Subsidies

The OECD defines subsidy as *"a result of a government action that confers an advantage on consumers or producers, in order to supplement their income or lower their costs"*. However, this definition ignores consequences for natural resources and does not cover subsidy as a result of inaction.

Environmentally harmful subsidies are a result of a government action or inaction that: *"confers an advantage on consumers or producers, in order to supplement their income or lower their costs, but in doing so, discriminates against sound environmental practices"*

### Subsidies that Distort Trade

Trade policies influence global trends in biodiversity. Provisions for trade in agriculture and fisheries (e.g. favourable treatments or preferential tariffs) can have a significant effect on land and resource-use patterns across exporting and importing countries. International trade agreements, combined with export-oriented national policies, can cause countries to focus on exporting natural resources at an unsustainable level. For example, the EU Fishing Agreements have led to exhaustion of resources by EU vessels outside the EU, leading to unsustainable use of natural resources in these countries.

**Fig.** Trade Policies Influence Global Trends in Biodiversity

Less obvious subsidies exist as accidental features of policies or lack of policies which means that the costs of damage to biodiversity and ecosystems are ignored. For example, water abstracted is rarely priced at its resource value, companies rarely pay for the value of genetic materials they build products on, and the cost of damage to forest or coastal areas is not generally paid for. This has already begun to change.

Although existing subsidies are well defended by vested interests, policy makers have recognized the importance of reforming them for environmental and economic reasons. Two avenues have proved to be promising. Subsidies can either be removed or reformed to promote environmentally friendly resource use – such as the changes to the agricultural subsidies in the United States of America and the European Union.

Subsidies can be replaced, using private resources to sustain financial flows for certain land-use practices, as in the example of landscape auction in the Netherlands. Landscapes are broken down into distinct elements such as a tree, a hedgerow or a pond. While the landowner still owns the item, people bid at auction to support the conservation of a specific element and thereby raise money for its preservation. Thus both farmers' incomes and the conservation of biodiversity can be assured without state subsidies.

**Rewarding Unrecognized Benefits, Penalizing Uncaptured Costs**

Getting prices right is a cardinal rule for good economics. Since most biodiversity and ecosystem benefits are in fact public goods that have no price, this can be done in two ways: instituting appropriate policies (which reward the preservation of the flow of these public goods and penalize their destruction), and encouraging appropriate markets (mainly "compliance markets" which attach tradable private values to the supply or use of these goods and create incentive structures to pay for them).

We highlight the example of payments for ecosystem services, and some nascent markets which could harness the power of supply and demand if appropriate infrastructure, incentives, financing and governance are provided.

## PAYMENTS FOR ECOSYSTEM SERVICES

Payments for ecosystem services (PES) can create demand, a necessary market force to correct an existing imbalance which harms biodiversity and stymies sustainable development. PES are payments for a service or the land use likely to secure that service. Governments are increasingly creating incentive programmes that support landowners who protect ecosystem services by compensating for lost revenues.

Payments are particularly valuable when land cannot be purchased and set aside for conservation, or where protected areas cannot be established. Payments can be international (IPES). One prominent example is the Clean Development Mechanism (CDM) that operates under the Kyoto Protocol. The Bali COP agreed to consider REDD (projects to Reduce Emissions from Deforestation and Forest Degradation) as part of the post- 2012 regime.

This is an important milestone, as it addresses the 18-20% of global greenhouse gas emissions from tropical deforestation and related land-use change. Preventing deforestation and creating and restoring forests can simultaneously protect biodiversity and ecosystem services as well as countering climate change. But significant funding is needed – possibly US$ 10 billion a year to achieve a substantial impact on deforestation rates – and there is still uncertainty about how to implement REDD and the scale of its ambitions.

Suitable financial mechanisms need to be designed to stimulate activity. One option is a market-based mechanism which would allow credits in avoided deforestation to be traded. The advantages of an early start with pilot schemes have to be weighed against the risks of leakage of deforestation pressure to neighbouring forests. REDD can significantly reduce greenhouse gas emissions at low cost, and at the same time help to conserve forests and their biodiversity.

However, potential risks from knock-on effects must be considered. REDD is unlikely to include support for ecosystem services other than carbon storage, and other services could be damaged by displaced deforestation pressures. For example, pressures to remove fuel wood and fodder from a degraded forest that has come under the purview of REDD could shift to a neighbouring forest area with healthier ecosystems and more biodiversity, which then would suffer. REDD could achieve reductions in emissions, but at the cost of biodiversity loss. PES can be substantial and support mainstream biodiversity policies.

The US government spends more than US$ 1.7 billion a year in direct payments to farmers for environmental protection. Payments under the Department of Agriculture's Environmental Quality Incentives Programme encourage sustainable use of irrigation, nutrients and fertilizers, integrated pest management and wildlife protection. Similarly, the European Union mechanism for promoting environmentally friendly agriculture and forestry is a major part of the EU Rural Development programmes, worth some EUR 4.5 billion annually.

In 2005, agri-environment schemes covered an area reaching 36.5 million hectares for EU-27, through 1.9 million contracts with farmers. PES can give communities the chance to improve their livelihoods through access to new markets. A feature of success is to combine "carrot and stick" by introducing protective legislation in conjunction with conservation incentives. This can be particularly significant for people in developing countries.

## EXTENDING THE "POLLUTER PAYS" PRINCIPLE

There is an increasing trend to use damage valuations to address degradation of biodiversity and ecosystem services. The polluter is frequently required to pay for damage caused, either through bearing the actual costs of clean-up and restoration projects, or through court-determined punitive damages. Significant examples include:

- *The Exxon-Valdez spill*: A 7,800-square-kilometre oil slick that still affects Alaska's fisheries, costing the polluter US$ 3.4 billion of penalties, clean-up costs and compensation.
- *Guadiamar River*: The main water source for the Doñana National Park salt marshes in Spain, which suffered from the devastating failure of a dam in Aznalcóllarmine, releasing toxicmuds for which cleanup and restoration efforts cost the Spanish authorities over EUR 150 million.

Such incidents have set large precedents for event-based recovery of costs. The "polluter pays" principle can be further extended through compliance markets, created so that cost externalities can be captured, securitized and capped in order to be traded amongst polluters who bear a market determined price for covering their pollution costs.

Fig. The Polluter

## CREATING NEW MARKETS

New markets are already forming which support and reward biodiversity and ecosystem services. Some of them have the potential to scale up. But to be successful, markets need appropriate institutional infrastructure, incentives, financing and governance – in short, investment. The state has traditionally been considered solely responsible for managing the public services of ecosystems, but it is now clear that markets can also contribute to this task, often without spending public money.

Market-based approaches can be flexible and cost-effective – a feature that traditional conservation policies often lack. There are, however, difficulties because "environmental services markets" may be imperfect, sometimes lacking depth and liquidity, and with limited competition. Price discovery is often not easy, since most ecosystem services are public services, delivered widely and often remotely in the form of positive externalities. In some cases transaction costs could offset potential gains.

Governments can help to remedy some of these shortcomings by providing an adequate institutional framework, for example by modifying liability rules, or capping resource use and issuing tradable permits to allow flexibility within the cap. The EU-ETS (European Union Emissions Trading Scheme for carbon credits) is a prime example of such a "compliance market". Governments can also facilitate private engagement to make ecosystem services visible, for example through labelling. Mechanisms and financial products have been developed to deal with environmental liabilities. Habitat and species banks are among the most innovative new instruments, providing tradable credits. Markets for products that are produced sustainably permit consumers to express their preferences for biodiversity and ecosystem protection in terms that businesses understand.

Such markets are growing fast – markets for organic agriculture, certified food and timber products are growing three times faster than the average and the market for sustainably produced commodities could reach US$ 60 billion a year by 2010.

In South Africa's Cape Floral Kingdom – a biodiversity hotspot home to nearly 10,000 plant species – wine producers who commit to conserving at least 10% of their vineyard are awarded "championship status" which they can advertise on product labels.

They can also raise revenue from ecotourism since the "Green Mountain Eco Route" was established in 2005. Such certification and eco-labelling are popular market-based instruments, although possibly with less long-term potential than the banking and trading schemes described here.

Businesses will also invest in ecosystem service management even if there are no direct products or reputational advantages, if the risks to the business of losing ecosystem services and the expected benefits are high enough. This makes a clear investment case on purely financial grounds for privately funded payments, as the Vittel-example demonstrates.

### Experience with Habitat Banking, Endangered Species Credits and Biobanking

In the United States of America, companies or individuals can buy environmental credits from Wetland Mitigation Banks to pay for degradation of wetland ecosystems due to agriculture or development activities. More than 400 banks had been approved by September 2005, almost three quarters of them sponsored by private entities, while in 2006 the trade of wetland bank credits reached an amount of US$ 350 million.

A biodiversity cap-and-trade system in the United States has created "endangered species credits", which can be used to offset a company's negative impacts on threatened species and their habitats. The market volume as of May 2005 was over US$ 40 million, with 930 transactions carried out and more than 44,600 hectares of endangered species habitat protected. In 2006, Australia began a pilot project in

New South Wales through the 2006 Bio Banking Bill to create incentives for protecting private land with high ecological value.

The project resulted in developers buying "biodiversity credits" to offset negative impacts on biodiversity. These credits can be created by enhancing and permanently protecting land.

### The Vittel Example

The Vittel mineral water company (Nestlé Waters) was concerned about nitrate contamination caused by agricultural intensification so it began to pay farmers within its catchment to make their practices more sustainable. A key element of success was that Vittel gained the farmers' trust and maintained their income levels by providing them with sufficiently large payments. It also financed any required technological changes, meaning that farmers were not out of pocket. The company worked intensively with farmers to identify suitable alternative practices and mutually acceptable incentives.

## PANAMA CANAL REFORESTATION

Insurance firms and major shipping companies are financing a 25-year project to restore forest ecosystems along the 80-kilometre length of the Panama Canal. The Canal is the preferred shipping route between the Atlantic and Pacific Oceans, with more than 14,000 vessels passing through in 2007. But its operation is becoming increasingly affected by floods, erratic water supply and heavy silting as a result of deforestation of the surrounding land. The costs of maintaining the canal are rising, and there is a growing risk that it will have to close.

Shipping companies faced increasing insurance premiums until Forest Re – a specialist insurance entity focused on forest risks – convinced them to fund ecosystem restoration. The advantages are less erosion and a more controlled flow of freshwater to the canal, which reduces insurance risk so that shippers enjoy lower premiums.

### Sharing the Benefits of Conservation

A better understanding of the economics of ecosystem services is crucial to safeguarding and extending protected areas, showing how to realise and share their value with local communities without jeopardizing their biodiversity benefits. More than 11% of the Earth's land surface is already legally protected thanks to a loose network of more than 100,000 protected areas, which together contain most types of terrestrial biodiversity.

The EU's Natura 2000 network is one example, accounting for around 20% of the EU-27 member territory. But the protected area network is not complete and those that exist are under threat from a lack of both funding and political support. Importantly in the context of our work, protected areas face financial pressure because of the potential to make money from timber, meat, biofuels and other resources.

The economic values of conservation need to be better understood and made more explicit. Valuation can help to inform policy choices on creating or maintaining protected areas. Examples like the Gabcíkovo-Nagymaros Barrage System in Hungary show that if the value of biodiversity is measured against the benefits of large development projects, the chance of protecting sensitive areas increases.

In this specific example, analysis showed that the natural capital involved far outweighed the benefit of the proposed dam project, which would have caused tremendous adverse impacts on biodiversity in the Szigetköz wetland areas.

Local communities are the first to bear the costs of biodiversity loss. They should share the benefits of conservation. Local communities as well as local governments typically look to achieve growth and economic development by attracting more people and businesses, promoting construction and infrastructure development. They may see protected areas as barriers to development, particularly where land is scarce and its use limited.

The resulting costs of limiting land use are borne locally, but the benefits are likely to reach far beyond municipal borders. This mismatch needs to be corrected, ideally by participation in the revenues from protected areas, as in Uganda. The costs of community-based conservation, such as livestock and crop losses, can be significant and need to be managed by communities, forest conservators and NGOs.

Inadequate compensation for losses is a common refrain, although there are other recent examples of success where a survey of local residents concluded that the socio-economic benefits outweighed the costs. Where the benefits are less direct than in the example above from Uganda, tax transfers between central, regional and local governments can provide local revenue that represents a share of ecosystem benefits. Brazil also demonstrates how this kind of financing works.

Protected areas in the state of Paraná have been valued in intergovernmental payments to municipalities since 1992. Quality indicators which determine payments take account of conservation goals achieved. As a result, the number of protected areas increased and their quality improved. Similar models have been developed in 12 of the 27 Brazilian states and others are considering this approach.

## Protected Areas in Uganda

Since 1995 Ugandan legislation places the management of natural resources in the hands of local authorities. Consequently the Ugandan Wildlife Authority (UWA) disburses 20%of all revenues from protected area (PA) tourism to the local communities neighbouring the PAs. This percentage has been fixed without a precise picture of PA economics, but even a rough approximation of costs and benefits allows local livelihoods to be enhanced while preserving biodiversity. Of course such a benefit sharing regime only works in the long run if it actually compensates for the use-restrictions that PAs imply for local communities. Thus, knowing better the costs and benefits involved will allow reconciling ongoing biodiversity conservation and enhancing rural livelihoods.

Some protected areas placed under the "Revenue Sharing Programme" of the Ugandan Wildlife Authority

- Bwindi Impenetrable National Park
- Mgahinga Gorilla National Park
- Lake Mburo National Park
- Queen Elizabeth National Park
- Rwenzori Mountains National Park
- Kibaale National Park
- Semliki National Park
- Murchison Falls National Park
- Mount Elgon National Park

**Table. Population Trends of Selected Species in Lake Mburo National Park**

| Species | 1999 | 2002 | 2003 | 2004 | 2006 |
|---|---|---|---|---|---|
| Zebra | 2,249 | 2,665 | 2,345 | 4,280 | 5,986 |
| Buffalo | 486 | 132 | 1,259 | 946 | 1,115 |
| Waterbuck | 598 | 396 | 899 | 548 | 1,072 |
| Hippo | 303 | 97 | 272 | 213 | 357 |
| Impala | 1,595 | 2,956 | 2,374 | 3,300 | 4,705 |

In Europe, Portugal has led the use of intergovernmental fiscal transfers to municipalities for Natura 2000 areas that relate to the EU Habitats and Birds Directives. The costs of loss and degradation relate to how much local communities depend on biodiversity and ecosystem services. Many indigenous communities are utterly dependent for survival on their local resources.

Especially in such cases, "community-conserved areas" based on traditionally sustainable resource-use systems are a further alternative and can be more effective than conventional protected areas. They could have governance structures that are adapted to local needs as well as the local skills and knowledge available. Valuing and sharing the benefits of biodiversity and ecosystem services can thus help biodiversity protection policies to better address the needs of local communities.

If benefits mainly occur beyond the local level, transfers can reward communities' efforts and help them find the resources needed for the protection of biodiversity and the provision of ecosystems services.

## WHAT ECOSYSTEM AND BIODIVERSITY ECONOMICS CAN OFFER TO PROTECTED AREAS

*A better understanding of biodiversity economics will help to*:

- *Create cash flow*: protected areas' chronic funding shortages totalled US$ 38.5 billion in 2001. Quantifying the financial and non-financial benefits of ecosystems is key to tapping private funding and generating income for protected areas by realizing payments for ecosystem services.
- *Gain political support*: clarity about the economic benefits of maintaining ecosystem services could increase political support to match that typically achieved for sectors such as agriculture, industrial development and regional planning.
- *Improve policy making*: introducing values for biodiversity and ecosystem services will support better policy decisions about land use, based on quantifying the effects of decisions and allowing evaluation of trade-offs between options, for example levels of grazing or wood extraction.
- *Improve the governance structures*: protected areas are often managed according to blueprints without taking into account the distribution of relevant competences and the concerns of those most affected by protection. A better understanding of the costs and benefits of conservation and use of biodiversity can help to improve the distribution of responsibilities in management.

## MEASURING WHAT WE MANAGE: METRICS FOR SUSTAINABILITY

Our economic compass is defective because of unaccounted externalities at every level – national, corporate and individual. Here we summarize work in progress to correct this failure, and describe how we can contribute in Phase II. The inadequacy of national accounting has been recognized for at least 40 years. It is now essential to aim "beyond GDP" as inadequate metrics have cost us dearly in terms of unsustainable growth, degraded ecosystems, lost biodiversity, and even reduced per-capita human welfare, especially in developing countries.

In November 2007, the European Commission, the European Parliament, the Club of Rome, WWF and the OECD held a major conference in Brussels called Beyond GDP. It was attended by 650 policy makers and opinion leaders from all over the world. It focused on the need for more than GDP as a measure of what society values, highlighted by the fact that devastation by events such as Hurricane Katrina and the Asian tsunami show up as increases in GDP despite the human tragedies and property losses.

The conference consensus was that we needed to add environmental and social measures to the existing GDP metric. Targeting solely for classical GDP growth may not be of much help with many of our pressing problems. For example, it may not be able to solve persistent poverty in Africa and Asia, nor equip us to tackle climate change and unsustainable development. The call for action comes not just from policy makers and experts, but also from the public.

In a survey on measures of progress beyond GDP, three-quarters of those asked (in 10 countries including Australia, Brazil, Canada, France, Germany and Russia) concluded that governments should *"look beyond economics and include health, social and environmental statistics in measuring national progress"*. The widely used System of National Accounts (SNA) does not recognize many significant externalities in the areas of natural resources, health and education.

This means that desirable improvements in health and education are accounted for as expenses instead of investments. Valuable ecosystem services which are income sources are not recognized at all and deforestation is not recorded as a form of depreciation. Managing improvements in health, education and quality of environment without a formal framework to value them financially can be a frustrating exercise. Sub-optimal policy choices and trade-offs are likely in the absence of a "sustainability yardstick".

The publication of a "genuine savings" indicator for many years by the World Bank showed that adding to the GDP metric was indeed possible at a global level. However, the usefulness of this metric was limited by its need to capture minimum standards in data collected across all countries, thus limiting the range of natural capital adjustments which could be included in computing genuine savings.

Developing amore inclusive national income and wealth accounting (NIWA) metric should be a priority, especially for countries most prone to ecosystem and biodiversity losses. It could make the difference between a viable and sustainable economic trajectory and one which spells disaster not just for developing countries but for us all. The United Nations' System of Integrated Environmental and Economic Accounting can be a starting point for preparing holistic national income and wealth accounts that reflect externalities in the areas of natural resources, health and education.

At present, few countries produce holistic national income statistics on this basis, and there is no comparability because different areas are covered, different externalities captured, and there are varying degrees of granularity. A revision of the SNA 1993 is currently being finalized by the UN Statistical Commission, involving many key organizations including UNEP, the World Bank, IMF, OECD, European Commission and statistical offices around the world. We understand that an important component of the SNA revision is the recognition of an enhanced version of the SEEA as a standard.

The ongoing SEEA revision process, initiated by the UN Committee of Experts on Environmental-Economics Accounting (UNCEEA) Bureau, is a timely and necessary step for national income metrics to progress "beyond GDP". We believe that ecosystems, biodiversity and their valuation deserve particularly close attention. It is very important that the development of ecosystem/biodiversity accounting in physical and monetary terms is promoted as a key early priority of the ongoing SEEA revision, building on the work of EEA and others.

At the corporate level, too, there is gradual recognition of the need to redefine corporate success, and enhance performance measurement and reporting to reflect a broader vision for the corporation than just an optimizer of financial capital for its shareholders. "Triple bottom line" and sustainability reporting disciplines are being followed by an increasing number of corporations. The Global Reporting Initiative (GRI) has issued detailed guidelines on sustainability reporting.

The Carbon Disclosure Project has been successful in eliciting voluntary disclosure from an increasing number of corporations and countries year after year. All these initiatives are, however, based on voluntary disclosure, and are not followed widely enough to be considered market standards.

In Phase II we will reach out to organizations involved in redefining corporate performance metrics and reporting standards as we intend to evolve valuation guidance on corporations' use of natural capital, including carbon footprint measurement. Consumers are a major source of pressures to convert natural ecosystems to other land uses, especially through demand for food. Different kinds of foods have dramatically different ecological footprints.

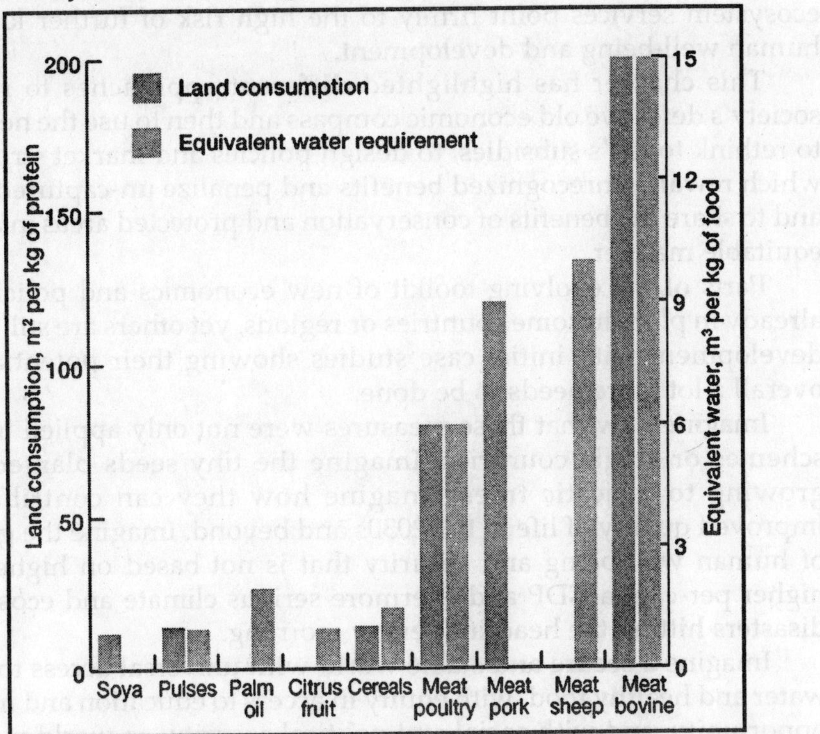


It is difficult for consumers to incorporate these factors in purchasing choices unless the products they buy – especially food – clearly disclose their ecological footprint at the point of sale. A credible standard methodology is a basic prerequisite, which we shall explore further with end-user groups in Phase II. The goal is to identify or evolve standard metrics for consumer footprint (in terms of land, water and energy use) which are based on sound ecology and economics simple enough to understand and to be implemented by retailers.

## IMAGINING A NEW WORLD

It is gradually becoming accepted that healthy ecosystems maintaining high levels of biodiversity are more resilient to external pressure and consequently better able to sustain the delivery of ecosystem services to human society. Countries and more and more companies and citizens want to know and understand the reality of the costs of using the Earth's natural capital and the consequences of policies on the resilience and sustainability of ecosystems. We still face many gaps in knowledge on the status and trends of biodiversity and the drivers and pressures that contribute to its loss, but the scenarios we have outlined on the projected loss of biodiversity, ecosystems and ecosystem services point firmly to the high risk of further losses to human well-being and development.

This chapter has highlighted different approaches to replace society's defective old economic compass and then to use the new one: to rethink today's subsidies, to design policies and market structures which reward unrecognized benefits and penalize un-captured costs, and to share the benefits of conservation and protected areas in a more equitable manner.

Parts of the evolving toolkit of new economics and policies are already in place in some countries or regions, yet others are still under development with initial case studies showing their potential, but overall a lot more needs to be done.

Imagine now that these measures were not only applied in pilot schemes or single countries. Imagine the tiny seeds planted now growing to majestic trees. Imagine how they can contribute to improved quality of life in the 2030s and beyond. Imagine the growth of human well-being and security that is not based on higher and higher per-capita GDP and evermore serious climate and ecosystem disasters hitting the headlines every morning.

Imagine a secure and stable world with universal access to clean water and healthy food, with equity in access to education and income opportunity, and with social and political security – a world meeting and even going beyond the Millennium Development Goals. Biodiversity and ecosystem services are now recognized as vital infrastructure to achieve human welfare and wellbeing.

We are convinced that The Economics of Ecosystems and Biodiversity, if used with careful consideration of the underlying ethical choices, can make decisive contributions towards safeguarding biodiversity and ecosystem services and improving well-being for us and for generations to come.

## Chapter 8

# Biodiversity and Livelihood

Biodiversity provides the basis for ecosystems and the services they provide, upon which all people fundamentally depend. The following are the main messages of this chapter: *People rely on biodiversity in their daily lives, often without realizing it.* Biodiversity contributes to many aspects of people's livelihoods and well-being, providing products, such as food and fibres, whose values are widely recognized. However, biodiversity underpins a much wider range of services, many of which are currently undervalued.

The bacteria and microbes that transform waste into usable products, insects that pollinate crops and flowers, coral reefs and mangroves that protect coastlines, and the biologically-rich landscapes and seascapes that provide enjoyment are only a few. Although much more remains to be understood about the relationships between biodiversity and ecosystem services, it is well established that if the products and services that are provided by biodiversity are not managed effectively, future options will become ever more restricted, for rich and poor people alike.

However, poor people tend to be the most directly affected by the deterioration or loss of ecosystem services, as they are the most dependent on local ecosystems, and often live in places most vulnerable to ecosystem change. Current losses of biodiversity are restricting future development options. Ecosystems are being transformed, and, in some cases, irreversibly degraded, a large number of species have gone extinct in recent history or are threatened with extinction, reductions in populations are widespread and genetic diversity is widely considered to be in decline.

It is well established that changes to biodiversity currently underway on land and in the world's fresh and marine waters are more rapid than at any time in human history, and have led to a degradation in many of the world's ecosystem services. Reducing the rate of loss of biodiversity, and ensuring that decisions made incorporate the full values of goods-and-services provided by biodiversity will contribute substantially towards achieving sustainable development as described in the report of the World Commission on Environment and Development (Brundtland Commission report).

- Biodiversity plays a critical role in providing livelihood security for people. It is particularly important for the livelihoods of the rural poor, and for regulating local environmental conditions. Functioning ecosystems are crucial as buffers against extreme climate events, as carbon sinks, and as filters for waterborne and airborne pollutants.
- From the use of genetic resources to harnessing other ecosystem services, agriculture throughout the world is dependent on biodiversity. Agriculture is also the largest driver of genetic erosion, species loss and conversion of natural habitats. Meeting increasing global food needs will require one or both of two approaches: intensification and extensification. Intensification is based on higher or more efficient use of inputs, such as more efficient breeds and crops, agrochemicals, energy and water. Extensification requires converting increasing additional areas of land to cultivation. Both approaches have the potential to dramatically and negatively affect biodiversity. In addition, the loss of diversity in agricultural ecosystems may undermine the ecosystem services necessary to sustain agriculture, such as pollination and soil nutrient cycling.
- Many of the factors leading to the accelerating loss of biodiversity are linked to the increasing use of energy by society. Dependence on and growing requirements for energy are resulting in significant changes in species and ecosystems, as a result of the search for energy sources and of current energy use patterns. The consequences can be seen at all levels: locally, where the availability of traditional biomass energy is under threat, nationally, where energy prices affect government policies, and globally, where climate change driven by fossil-fuel use is changing species ranges and behaviour. The latter is likely to have very significant consequences for livelihoods, including changing patterns of human infectious disease distribution, and increased opportunities for invasive alien species.
- Human health is affected by changes in biodiversity and ecosystem services. Changes to the environment have altered disease patterns and human exposure to disease outbreaks. In addition, current patterns of farming, based on high

resource inputs (such as water and fertilizers) and agricultural intensification, are putting great strains on ecosystems, contributing to nutritional imbalances and reduced access to wild foods.

- Human societies everywhere have depended on biodiversity for cultural identity, spirituality, inspiration, aesthetic enjoyment and recreation. Culture can also play a key role in the conservation and sustainable use of biodiversity. Loss of biodiversity affects both material and non-material human well-being. Both the continued loss of biodiversity and the disruption of cultural integrity represent obstacles towards the attainment of the Millennium Development Goals (MDGs).

Biodiversity loss continues because current policies and economic systems do not incorporate the values of biodiversity effectively in either the political or the market systems, and many current policies are not fully implemented. Although many losses of biodiversity, including the degradation of ecosystems, are slow or gradual, they can lead to sudden and dramatic declines in the capacity of biodiversity to contribute to human wellbeing. Modern societies can continue to develop without further loss of biodiversity only if market and policy failures are rectified.

These failures include perverse production subsidies, undervaluation of biological resources, failure to internalize environmental costs into prices and failure to appreciate global values at the local level. Reducing the rate of biodiversity loss by 2010 or beyond will require multiple and mutually supportive policies of conservation, sustainable use and the effective recognition of value for the benefits derived from the wide variety of life on Earth. Some such policies are already in place at local, national and international scales, but their full implementation remains elusive.

## IMPORTANCE OF BIODIVERSITY

The understanding of the importance of biodiversity has developed in the 20 years since the report of the World Commission on Environment and Development (Brundtland Commission). There is increased recognition that people are part of, not separate from, the ecosystems in which they live, and are affected by changes in ecosystems, populations of species and genetic changes. Along with human health and wealth, human security and culture are strongly affected by changes in biodiversity, and associated impacts on ecosystem services.

As the basis for all ecosystem services, and the foundation for truly sustainable development, biodiversity plays fundamental roles in maintaining and enhancing the well-being of the world's more than 6.7 billion people, rich and poor, rural and urban alike. Biodiversity comprises much of the renewable natural capital on which livelihoods and development are grounded. However, ongoing, and in many cases, accelerating declines and losses in biodiversity over the past 20 years have decreased the capacity of many ecosystems to provide services, and have had profound negative impacts on opportunities for sustainable development around the planet.

These impacts are particularly pronounced in the developing world, in large part due to the patterns of consumption and trade in the industrial world, which themselves are not sustainable. If future concerns are not taken into account, and the products and services provided by biodiversity are not managed effectively, future options become limited or are eliminated, for rich and poor people alike.

While technological alternatives to some of the services provided by biodiversity are available, they are typically more costly, compared to the benefits derived from well-managed ecosystems.

Biodiversity loss particularly affects the poor, who are most directly dependent on ecosystem services at the local scale, and are unable to pay for alternatives.

Although the private, more restricted, financial benefits of activities that result in the loss of biodiversity, such as the conversion of mangroves to aquaculture enterprises, are usually high, they often externalize many of the social and environmental costs. The overall benefits are frequently considerably less than the societal, more distributed, benefits that are lost along with the biodiversity, but for which the monetary value is often not known.

For example, the loss of mangrove ecosystems contributes to declining fisheries, timber and fuel, the reduction of storm protection, and increased vulnerability to the impacts of extreme events. In addition to the values of biodiversity for the supply of particular ecosystem services, biodiversity also has intrinsic value, independent from its functions and other benefits to people.

The challenge is to balance the cultural, economic, social and environmental values so that the biodiversity of today is conserved and used in a manner that will allow it to be available for and to sustain the generations of the future.

Biodiversity management and policies have an impact upon all sectors of society, and have strong cross-cultural and cross-boundary implications. Policies relating to issues such as trade, transport, development, security, health care and education all have impacts on biodiversity. Discussions on access and benefit sharing relating to genetic resources, one of the provisions of the UN Convention on Biological Diversity (CBD), show that understanding the full value of biodiversity is not simple.

In addition to the gaps remaining in the understanding of biodiversity and ecosystem functioning, each individual stakeholder may hold different values for the same attribute of biodiversity. Building a fuller understanding of these values will require considerable additional research, and increasingly comprehensive, interdisciplinary and quantified assessments of the benefits that biodiversity provides to people's health, wealth and security.

## Life on Earth

Biodiversity is the variety of life on Earth. It includes diversity at the genetic level, such as that between individuals in a population or between plant varieties, the diversity of species, and the diversity of ecosystems and habitats. Biodiversity encompasses more than just variation in appearance and composition. It includes diversity in abundance (such as the number of genes, individuals, populations or habitats in a particular location), distribution (across locations and through time) and in behaviour, including interactions among the components of biodiversity, such as between pollinator species and plants, or between predators and prey.

Biodiversity also incorporates human cultural diversity, which can be affected by the same drivers as biodiversity, and which has impacts on the diversity of genes, other species and ecosystems. Biodiversity has evolved over the last 3.8 billion years or so of the planet's approximately 5 billion-year history. Although five major extinction events have been recorded over this period, the large number and variety of genes, species and ecosystems in existence today are the ones with which human societies have developed, and on which people depend.

The relationships among biodiversity and the five main themes assessed in this chapter – livelihood security, agriculture, energy, health and culture – clearly demonstrate the importance of biodiversity to these aspects of human well-being. Biodiversity forms the basis of agriculture, and enables the production of foods, both wild and cultivated, contributing to the health and nutrition of all people. Genetic resources have enabled past and current crop and livestock improvements, and will enable future ones, and allow for flexibility according to market demand and adaptation according to changing environmental conditions. Wild biodiversity is perhaps of greatest direct importance to the one billion people around the world who live a subsistence lifestyle.

The decline of this diversity has considerable implications for their health, culture and livelihoods. Supporting services, such as nutrient cycling and soil formation, and regulating services, such as pest and disease control, flood regulation and pollination, underpin successful agricultural systems, and contribute to livelihood security.

Cultural ecosystem services are being increasingly recognized as key determinants of human well-being, including through the maintenance of cultural traditions, cultural identity and spirituality. Among the wide range of other benefits from biodiversity, it has enabled the production of energy from biomass and fossil fuels. Such use of biodiversity has brought tremendous benefit to many people, but has had some significant negative knock-on effects in the form of human-induced climate change, and habitat conversion.

These trade-offs, inherent in so much of biodiversity use, are becoming increasingly apparent, as there are greater demands for ecosystem services. People directly use only a very small percentage of biodiversity. Agriculture reduces diversity to increase productivity for a component of biodiversity of particular interest. However, people rely indirectly on a much larger amount of biodiversity without realizing it. There are bacteria and microbes that transform waste into usable products, insects that pollinate crops and flowers, and biologically diverse landscapes that provide inspiration and enjoyment around the world.

Such ecosystem services, or the benefits derived from biodiversity, are ultimately dependent on functioning ecosystems. However, the amount of biodiversity required to enable ecosystems to function effectively varies enormously, and how much biodiversity is needed for the sustainable supply of ecosystem services in the present, and into the future, remains largely unknown. Despite the critical need for more effective conservation and sustainable use, the loss of biodiversity continues, and in many areas is currently increasing in magnitude. Rates of species extinction are 100 times higher than the baseline rate shown by the fossil record.

The losses are due to a range of pressures, including land-use change and habitat degradation, overexploitation of resources, pollution and the spread of invasive alien species. These pressures are themselves driven by a range of socio-economic drivers, chiefly the growing human population and associated increases in global consumption of resources and energy, and the inequity associated with high levels of per capita consumption in developed countries. Responses to the continuing loss of biodiversity are varied, and include further designation of protected areas, and, increasingly, the improved management for biodiversity in production landscapes and seascapes.

There are recent signs of an emerging consensus that biodiversity conservation and sustainable development are inextricably linked, as for example illustrated by the endorsement by the 2002 Johannesburg World Summit on Sustainable Development (WSSD) of the CBD's 2010 target, and its subsequent incorporation into the Millennium Development Goals.

**Value of biodiversity and ecosystem services**

The supply of ecosystem services depends on many attributes of biodiversity. The variety, quantity, quality, dynamics and distribution of biodiversity that is required to enable ecosystems to function, and the supplying benefits to people, vary between services. The roles of biodiversity in the supply of ecosystem services can be categorized as provisioning, regulating, cultural and supporting, and biodiversity may play multiple roles in the supply of these types of services.

For example, in agriculture, biodiversity is the basis for a provisioning service (food, fuel or fibre is the end product), a supporting service (such as micro-organisms cycling nutrients and soil formation), a regulatory service (such as through pollination), and potentially, a cultural service in terms of spiritual or aesthetic benefits, or cultural identity.

**Fig.** Honey bees (*Apis mellifera, Apis mellifica*) Provide Regulatory Services Through Pollination.

The contributions of biodiversity-dependent ecosystem services to national economies are substantial. The science of valuation of ecosystem services is new, and still developing basic conceptual and methodological rigour and agreement, but it has already been very instructive, since the value of such services is generally ignored or underestimated at decision and policy making levels.

Identifying economic values of ecosystem services, together with the notions of intrinsic value and other factors, will assist significantly in future decisions relating to trade-offs in ecosystem management.

*Value of*:

- Annual world fish catch – US$58 billion (provisioning service).
- Anti-cancer agents from marine organisms – up to US$1 billion/year (provisioning service).
- Global herbal medicine market – roughly US$43 billion in 2001 (provisioning service).
- Honeybees as pollinators for agriculture crops – US$2–8 billion/year (regulating service).
- Coral reefs for fisheries and tourism – US$30 billion/year.

*Cost of*:

Mangrove degradation in Pakistan – US$20 million in fishing losses, US$500 000 in timber losses, US$1.5 million in feed and pasture losses (regulating provisioning services).

Newfoundland cod fishery collapse – US$2 billion and tens of thousands of jobs (provisioning service).

Of those ecosystem services that have been assessed, about 60 per cent are degraded or used unsustainably, including fisheries, waste treatment and detoxification, water purification, natural hazard protection, regulation of air quality, regulation of regional and local climate, and erosion control. Most have been directly affected by an increase in demand for specific provisioning services, such as fisheries, wild meat, water, timber, fibre and fuel.

## GLOBAL OVERVIEW OF THE STATUS OF BIODIVERSITY

### Ecosystems

Ecosystems vary greatly in size and composition, ranging from a small community of microbes in a drop of water, to the entire Amazon rain forest. The very existence of people, and that of the millions of species with which the planet is shared, is dependent on the health of our ecosystems. People are putting increasing strain on the world's terrestrial and aquatic ecosystems. Despite the importance of ecosystems, they are being modified in extent and composition by people at an unprecedented rate, with little understanding of the implications this will have in terms of their ability to function and provide services in the future.

Figure depicts an analysis of the status of terrestrial ecosystems. For more than half of the world's 14 biomes, 20–50 per cent of their surface areas have already been converted to croplands. Tropical dry broadleaf forests have undergone the most rapid conversion since 1950, followed by temperate grasslands, flooded grasslands and savannahs. Approximately 50 per cent of inland water habitats are speculated to have been transformed for human use during the twentieth century.

Some 60 per cent of the world's major rivers have been fragmented by dams and diversions, reducing biodiversity as a result of flooding of habitats, disruption of flow patterns, isolation of animal populations and blocking of migration routes. River systems are also being significantly affected by water withdrawals, leaving some major rivers nearly or completely dry. In the marine realm, particularly threatened ecosystems include coral reefs and seamounts.

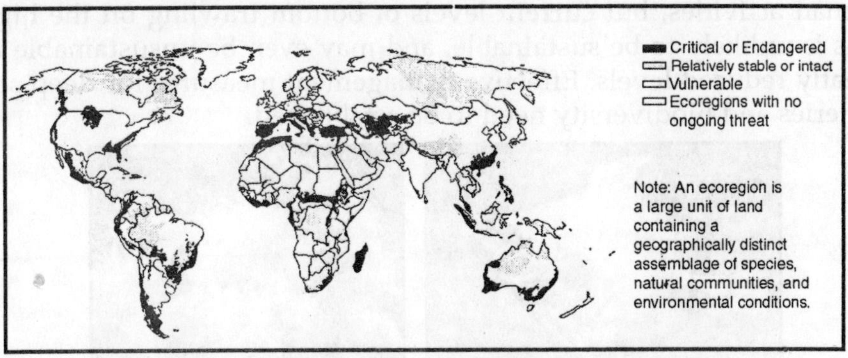


**Fig.** Status of Terrestrial Ecoregions

## DEEP-SEA BIODIVERSITY

The deep sea is increasingly recognized as a major reservoir of biodiversity, comparable to the biodiversity associated with tropical rain forests and shallow-water coral reefs. The wealth of diverse deepsea habitats – hydrothermal vents, cold seeps, seamounts, submarine canyons, abyssal plains, oceanic trenches and recently-discovered asphalt volcanoes – contain a vast array of unique ecosystems and endemic species. Although the magnitude of deep-sea diversity is not yet understood (only 0.0001 per cent of the deep seabed has been subject to biological investigations), it has been estimated that the number of species inhabiting the deep sea may be as high as 10 million.

It is believed that the deep seabed supports more species than all other marine environments. Marine biodiversity and ecosystems are threatened by pollution, shipping, military activities and climate change, but today fishing presents the greatest threat. The emergence of new fishing technologies and markets for deep-sea fish products has enabled fishing vessels to begin exploiting these diverse, but poorly understood deep-sea ecosystems. The greatest threat to biodiversity in the deep sea is bottom trawling.

This type of high seas fishing is most damaging to seamounts and the coldwater corals they sustain. These habitats are home for several commercial bottom-dwelling fish species. Seamounts are also important spawning and feeding grounds for species, such as marine mammals, sharks and tuna, which make them very attractive fishing grounds. The long life cycles and slow sexual maturation of deep-sea fish make them particularly vulnerable to large-scale fishing activities.

The lack of data on deep-sea ecosystems and associated biodiversity makes it difficult to predict and control the impacts of human activities, but current levels of bottom trawling on the high seas is unlikely to be sustainable, and may even be unsustainable at greatly reduced levels. Effective management measures for deep-sea fisheries and biodiversity need to be established.

Examples of species inhabiting the deep sea. False boarfish, Neocytlus helgae (left) and cold water coral, Lophelia (right).

Conservation of marine ecosystems has recently extended to the deep sea with the designation in 2003 of the Juan de Fuca Ridge system and associated Endeavour Hydrothermal Vents (2 250 metres deep and 250 kilometres south of Vancouver Island, Canada) as a marine protected area. There are several mechanisms to conserve deep seas, such as the 1982 UN Convention of the Law of the Sea (UNCLOS), 1995 UN Fish Stocks Agreement (UNFSA),

International Seabed Authority (ISA), 1992 Convention on Biological Diversity (CBD) and the 1973 Convention on Trade in Endangered Species (CITES). However, these mechanisms need more effective implementation if deep-sea ecosystems are to be conserved and sustainably used.

The seafloor off Northwest Australia showing dense populations of corals and sponges before trawling (left) and after trawling (right).

The fragmentation of ecosystems is increasingly affecting species, particularly migratory species that need a contiguous network of sites for their migratory journeys, species that rely on particular microhabitats and those that require multiple types of habitats during different life cycle stages.

## SPECIES

Although about 2 million species have been described, the total number of species range between 5 and 30 million. Much of this uncertainty relates to the most species-rich groups such as invertebrates. Current documented rates of extinction are estimated to be roughly 100 times higher than typical rates in the fossil record. Although conservation success in the recovery of several threatened species has been noted, and a few species that were presumed extinct have been rediscovered, it is feasible that extinction rates will increase to the order of 1 000–10 000 times background rates over the coming decades.

Fewer than 10 per cent of the world's described species have thus far been assessed to determine their conservation status. Of these, over 16 000 species have been identified as threatened with extinction. Of the major vertebrate groups that have been comprehensively assessed, over 30 per cent of amphibians, 23 per cent of mammals and 12 per cent of birds are threatened. To understand trends in extinction risk, the conservation status of an entire species group must be assessed at regular intervals.

Currently, this information is only available for birds and amphibians, both of which indicate a continuing increase in the risk of extinction from the 1980s to 2004. The threat status of species is not evenly distributed. Tropical moist forests contain by far the highest number of threatened species, followed by tropical dry forests, montane grasslands and dry shrublands. The distribution of threatened species in freshwater habitats is poorly known, but regional assessments from the United States, the Mediterranean Basin and elsewhere indicate that freshwater species are, in general, at much greater risk of extinction than terrestrial taxa.

Fisheries have also been greatly depleted, with 75 per cent of the world's fish stocks fully or overexploited. The Living Planet Index measures trends in the abundance of species for which data is available around the world. Despite the fact that invertebrates comprise the vast majority of species, trend indices for invertebrate groups only exist for a very small number of species groups, such as butterflies in Europe. The existing limited information suggests that invertebrate and vertebrate population declines may be similar, but further studies are required.

Fig. Invertebrates, Including Butterflies, Comprise the Vast Majority of Species. Genes

Genetic diversity provides the basis for adaptation, allowing living organisms to respond to natural selection, and adapt to their environment. Genes therefore play a strong role in the resilience of biodiversity to global changes, such as climate change or novel diseases. Genes also provide direct benefits to people, such as the genetic material needed for improving yield and disease resistance of crops or for developing medicines and other products.

Over the past two decades, many of the world's most important agricultural crops have lost genetic diversity due to changes in agricultural practices. The continued loss of genetic diversity of such crops may have major implications on food security. The amount or rate of loss of genetic diversity is poorly known, but inferences can be made from documented extinctions and population declines, which suggest that substantial genetic loss is occurring.

Global responses to curb biodiversity loss In 2002, parties to the CBD committed themselves to actions to "achieve, by 2010, a significant reduction of the current rate of biodiversity loss at the global, regional and national levels as a contribution to poverty alleviation and to the benefit of all life on earth". Setting this target has helped to highlight the need for improved biodiversity indicators, capable of measuring trends in a range of aspects of global biodiversity.

It has also helped to galvanize the scientific community to try to develop indicators capable of measuring trends in the various aspects or levels of biodiversity. Figure provides a sample of global biodiversity indicators that will be used to measure progress towards the 2010 target. They measure trends in vertebrate populations, extinction risks for birds, global consumption and the establishment of protected areas.

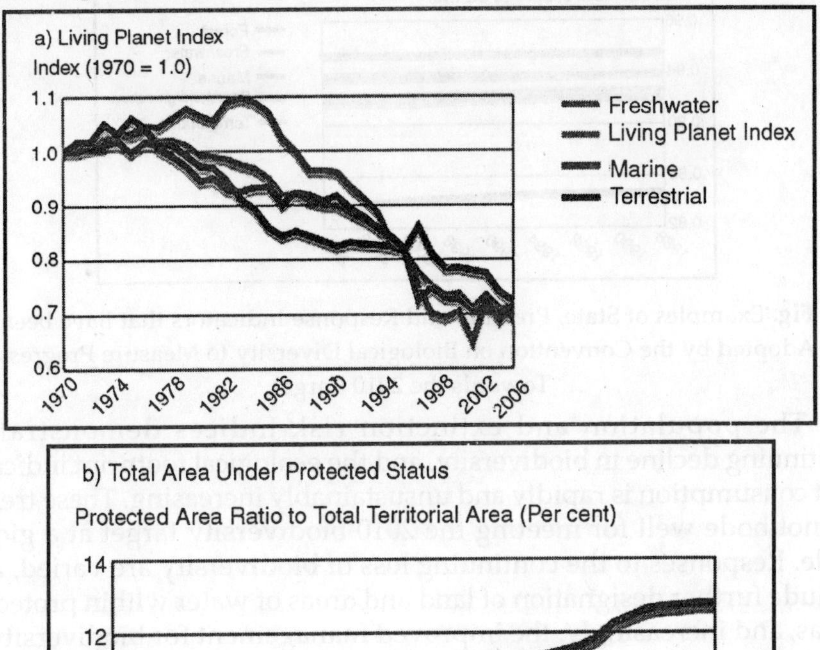


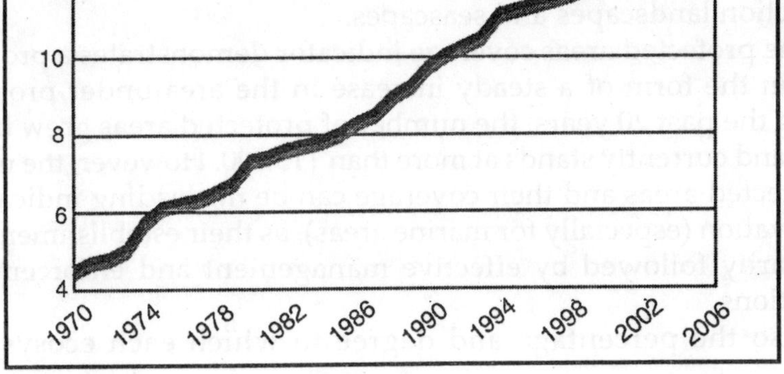

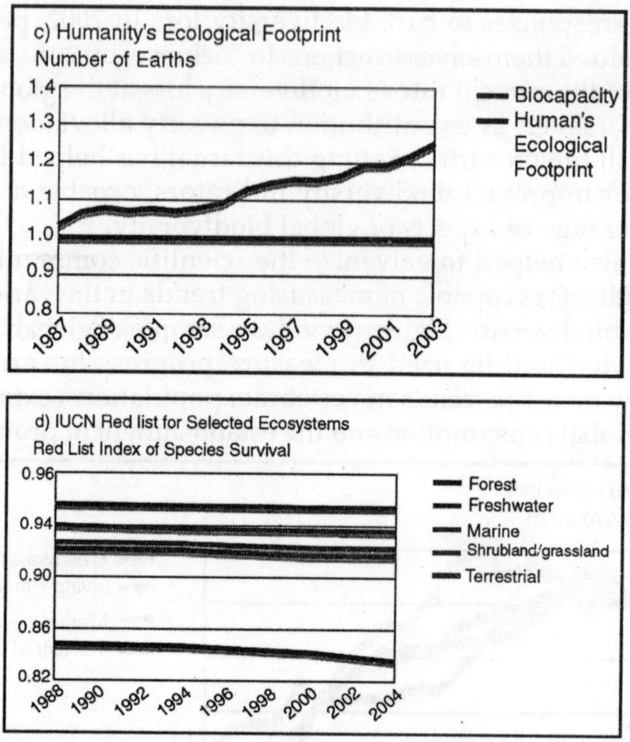


**Fig.** Examples of State, Pressure and Response Indicators that have been Adopted by the Convention on Biological Diversity to Measure Progress Towards the 2010 Target.

The population and extinction risk indices demonstrate a continuing decline in biodiversity, and the ecological footprint indicates that consumption is rapidly and unsustainably increasing. These trends do not bode well for meeting the 2010 biodiversity target at a global scale. Responses to the continuing loss of biodiversity are varied, and include further designation of land and areas of water within protected areas, and increasingly, the improved management for biodiversity in production landscapes and seascapes.

The protected areas coverage indicator demonstrates a promising trend in the form of a steady increase in the area under protection. During the past 20 years, the number of protected areas grew by over 22 000 and currently stands at more than 115 000. However, the number of protected areas and their coverage can be misleading indicators of conservation (especially for marine areas), as their establishment is not necessarily followed by effective management and enforcement of regulations.

Also the percentage and degree to which each ecosystem is

protected varies greatly. Roughly 12 per cent of the world's land surface is included within some kind of protected area, but less than one per cent of the world's marine ecosystems are protected, with the Great Barrier Reef and the northwestern Hawaiian islands making up one-third the area of all marine protected areas. In addition to ensuring the effective management of protected areas, emphasis will increasingly need to be placed on the conservation of biodiversity outside protected areas, and in conjunction with other land uses if the rate of loss of biodiversity is to be reduced.

The establishment of new policies and processes at all scales, the reemergence of sustainable agricultural practices, the further development of collaboration among sectors, including corporate partnerships between conservation organizations and extractive industries, and the mainstreaming of biodiversity issues into all areas of decision making, will all contribute to a more secure future for biodiversity, and for sustainable development. Over the last 20 years environmental issues have increasingly been recognized as important in the development sector at a global scale.

The commitment by parties to the CBD to achieve a significant reduction in the rate of biodiversity loss by 2010 as a contribution to poverty alleviation and the benefit of all life on Earth, the endorsement by the 2002 Johannesburg World Summit on Sustainable Development (WSSD) of the CBD's 2010 target, and the incorporation of the 2010 biodiversity target into the Millennium Development Goals as a new target under Goal 7 on environmental sustainability are some examples.

A framework for action was proposed at WSSD to implement sustainable development policies, which covered five key areas (water, energy, health, agriculture and biodiversity), This "WEHAB" framework provided a focus, and confirmed the recognition of biodiversity as a key component of the sustainable development agenda.

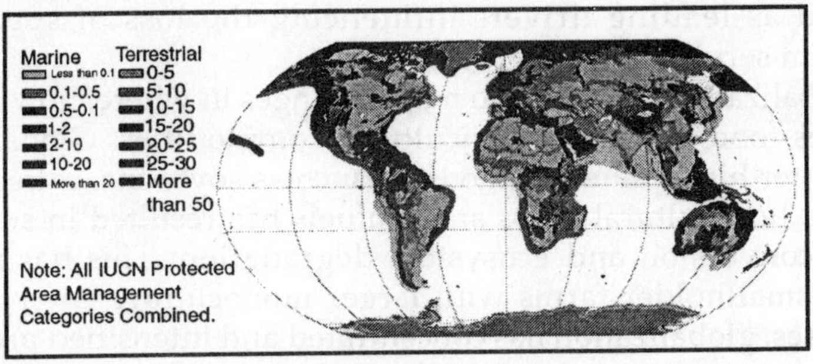


**Fig.** Degree of Protection of Terrestrial Ecoregions and L0arge Marine Ecosystems (per cent).

## DRIVERS OF CHANGE AND PRESSURES

Currently, population growth and patterns of consumption, which lead to increased demand for ecosystem services and energy, are the most important drivers affecting biodiversity. These drivers result in pressures that have direct impacts on ecosystems, species and genetic resources.

**Fig.** Deforestation

Human activities cause changes in both the living and non-living components of ecosystems, and these pressures have increased dramatically over the past few decades. Drivers and pressures seldom act in isolation. They tend to interact in synergistic ways, and their impacts on biodiversity are more than the sum of the effects of the individual drivers and pressures themselves.

Additionally, the interaction shows considerable regional variation. Drivers and pressures act at different temporal and spatial scales. For example, sediments from deforestation in the headwaters of the Orinoco River, deep in South America, have impacts far out in the Wider Caribbean Sea basin, changing the nutrient availability and turbidity of the waters. Since the Brundtland Commission report, the globalization of agriculture and inappropriate agricultural policies have emerged as leading drivers influencing the loss of species and ecosystem services.

Globalization is leading to major changes in where, how and who produces food and other agricultural commodities. Global market demand for high value commodities such as soybeans, coffee, cotton, oil palm, horticultural crops and biofuels has resulted in substantial habitat conversion and ecosystem degradation. This has replaced diverse smallholder farms with larger monoculture enterprises. In other cases, globalization has concentrated and intensified production on the most productive lands, reducing net deforestation rates.

Virtually all of the factors leading to the accelerating loss of biodiversity are linked to the development of and increasing demand for energy by society. Of particular importance are the high levels of per capita energy use in the developed world, and the potential growth in energy use in the large emerging economies. The rapid increase in demand for energy has profound impacts on biodiversity at two levels: impacts from the production and distribution of energy, and those resulting from the use of energy.

Exploration for hydrocarbons, pipeline construction, uranium and coal mining, hydroelectric dam construction, harvesting for fuel-wood and, increasingly, bio-fuel plantations can all lead to significant biodiversity loss, both on land and at sea. The widespread anthropogenic changes to the environment have altered patterns of human disease, and increased pressures on human well-being. The loss of genetic diversity, overcrowding and habitat fragmentation all increase susceptibility to disease outbreaks. Some ecosystem changes create new habitat niches for disease vectors, for example, increasing the risk of malaria in Africa and the Amazon Basin.

Trends in biodiversity over the next few decades will largely depend on human actions, especially those relating to land-use changes, energy production and conservation. These actions will, in turn, be affected by various factors including advances in our understanding of ecosystem services, development of viable alternatives to natural resources (especially fossil fuels), and the emphasis placed on the environment and conservation by developed and developing country governments alike.

Efforts made to predict the prospects for species-level biodiversity have indicated that extinctions are likely to continue at a pace well above the background rate, with as many as 3.5 per cent of the world's birds, and perhaps a greater proportion of amphibians and freshwater fish, being lost or committed to extinction by the middle of the century. Climate change is likely to play an increasing role in driving changes in biodiversity, with species' distributions and relative abundances shifting as their preferred climates move towards the poles and higher altitudes, leaving those endemic to polar and high mountain regions most at risk.

In addition, changes in the ranges of vector species may facilitate the spread of diseases affecting humans and other species, for example, malaria and the amphibian fungal disease, chytridiomycosis. Further pressure on biodiversity will result from the continuing increase in the global human population, which is predicted to reach 8 billion by 2025. All will require access to food and water, leading to an unavoidable increase in stresses on natural resources. The increased infrastructure required to support such a global population of more than 8 billion people will likely have particular effects on biodiversity in the future.

The increased need for agricultural production to feed the population will likely be met largely by commercial intensification, with negative consequences for the genetic diversity of agricultural crops and livestock. Extensification will also help to meet the need, with a predicted additional 120 million hectares required by 2030 in developing countries, including lands of high biodiversity value. Tropical forests are the terrestrial system likely to be the most affected by human actions in the first half of this century, largely through habitat conversion for agricultural expansion (including the growth of biofuel plantations). Ongoing fragmentation will result in the degradation of the largest remaining areas of speciesrich forest blocks in Amazonia and the Congo basin. Marine and coastal ecosystems are also expected to continue to be degraded, with existing impacts, such as fishing, eutrophication from terrestrial activities and coastal conversion for aquaculture, increasing. Large species, including top predators, will be particularly affected, with considerable declines and some extinctions likely. Changes, both positive and negative, in biodiversity trends over the next few decades are inevitable, yet the details of these changes are not yet set in stone.

Their magnitude can be somewhat reduced and mitigated by the further integration of biodiversity considerations into national policies, increasing corporate social responsibility activities and conservation actions. With commitment from governments, the private sector, scientific institutions and civil society, action can be taken to ensure progress towards the CBD 2010 target, the Millennium Development Goals, and beyond.

**Fig.** The *Telestes Polylepis,* a Critically Endangered Fresh Water Species Found in Croatia.

## ENVIRONMENTAL TRENDS AND RESPONSES

Biodiversity is closely linked to livelihood security, agriculture, energy, health and culture, the five themes analysed in this chapter. Of these themes, agriculture (in terms of food security) and energy were explicitly considered in the Brundtland Commission report, and with a focus on water and health, tie together the WEHAB framework for action arising from WSSD. These linkages are likely to emerge as the most critical in implementing actions that will result in truly sustainable development.

## LIVELIHOOD SECURITY

Ecosystems provide critical services Biodiversity contributes directly and indirectly to livelihood security. Functioning ecosystems are crucial buffers against extreme climate events, and act as carbon sinks and filters for water-borne and airborne pollutants. For example, the frequency of shallow landslides appears to be strongly related to vegetation cover, as roots play an important role in slope stability, and can give the soil mechanical support at shallow depth.

In coastal areas, mangroves and other wetlands are particularly effective in providing shoreline stability, reducing erosion, trapping sediments, toxins and nutrients, and acting as wind and wave breaks to buffer against storms. The role of inland wetlands in storing water and regulating stream-flow is both a function of their vegetative composition, which helps to maintain soil structure, and their characteristically gentle slopes.

**Table. Impacts on Biodiversity of Major Pressures and Associated Effects on Ecosystem Services and Human Well-being.**

| Pressures | Impacts on biodiversity | Potential implications for ecosystem services and human well-being |
|---|---|---|
| Habitat conversion | Decrease in natural habitat<br>Homogenization of species composition<br>Fragmentation of landscapes<br>Soil degradation | Increased agricultural production<br>Loss of water regulation potential<br>Reliance on fewer species<br>Decreased fisheries<br>Decreased coastal protection<br>Loss of traditional knowledge |
| Invasive alien species | Competition with and predation on native | Loss of traditionally available resources |

| | | |
|---|---|---|
| | **species**<br>**Changes in ecosystem function**<br>**Extinctions**<br>**Homogenization**<br>**Genetic contamination** | **Loss of potentially useful species**<br>**Losses in food production**<br>**Increased costs for agriculture,forestry, fisheries, water management and human health Disruption of water transport** |
| **Overexploitation** | **Extinctions and decreased populations**<br>**Alien species introduced after resource depletion**<br>**Homogenization and changes in ecosystem functioning** | **Decreased availability of resources**<br>**Decreased income earning potential**<br>**Increased environmental risk (decreased resilience)**<br>**Spread of diseases from animals to people** |
| **Climate change** | **Extinctions**<br>**Expansion or contraction of species ranges** | **Changes in resource availability**<br>**Spread of diseases to new** |
| **Changes in species** | **ranges** | |
| | **compositions and interactions** | **Changes in the characteristics of protected areas**<br>**Changes in resilience of ecosystems** |
| **Pollution** | **Higher mortality rates**<br>**Nutrient loading**<br>**Acidification** | **Decreased resilience of service**<br>**Decrease in productivity of service**<br>**Loss of coastal protection,** |
| | **with the degradation** | **of reefs and mangroves**<br>**Eutrophication, anoxic waterbodies**<br>**leading to loss of fisheries** |

Current trends in land degradation and habitat loss continue to contribute to reducing livelihood options while heightening risks. Changes in land management, particularly the replacement of fire-adapted systems with other forms of land cover, can increase the intensity and extent of fires, increasing the hazard to people. Land-use change also influences climate at local, regional and global scales.

Forests, shrub and grasslands, freshwater and coastal ecosystems provide critical sources of food and complementary sources of income. Fish and wild meat provide animal protein, while other forest resources provide dietary supplements.

These ecosystem goods act as critical safety nets for millions of rural poor. Traditionally, access rights and tenure arrangements for these public goods have evolved to enable equitable distribution of such extractive activities. More recently, due to increased population densities and the introduction of market models, access to these common property resources has been increasingly restricted, with resulting impacts on rural livelihoods. With reliable access to markets, the commercialization of many wild-harvested products can be extremely successful in contributing to sustaining rural livelihoods.

## CORAL REEFS IN THE CARIBBEAN

The global net value of coral reefs relating to fisheries, coastal protection, tourism and biodiversity, is estimated to total US$29.8 billion/year. However, nearly two-thirds of Caribbean coral reefs are reported to be threatened by human activities. The predominant pressure in the region is overfishing, which affects approximately 60 per cent of Caribbean reefs.

Other pressures include large quantities of dust originating from deserts in Africa, which are blown across the Atlantic Ocean and settle on reefs in the Caribbean, leading to significant coral mortality. It has been proposed that this phenomenon led to a coral bleaching event that began in 1987, correlating with one of the years of maximum dust flux into the Caribbean. Coral degradation has negative impacts on coastal communities, including the loss of fishing livelihoods, protein deficiencies, loss of tourism revenue and increased coastal erosion.

### Mangrove Restoration for Buffering Storm Surges in Viet Nam

In Viet Nam, tropical cyclones have caused a considerable loss of livelihood resources, particularly in coastal communities. Mangrove ecosystem rehabilitation along much of Viet Nam's coastline is an example of a cost-effective approach to improving coastal defences while generating local livelihoods. Since 1994, the Viet Nam National Chapter of the Red Cross has worked with local communities to plant and protect mangrove forests in northern Viet Nam.

Nearly 120 km$^2$ of mangroves have been planted, with substantial resulting benefits. Although planting and protecting the mangroves cost approximately US$1.1 million, it saved US$7.3 million/ year in dyke maintenance. During the devastating typhoon Wukong in 2000, project areas remained unharmed, while neighbouring provinces suffered huge losses in lives, property and livelihoods. The Viet Nam Red Cross has estimated that some 7 750 families have benefited from mangrove rehabilitation. Family members can now earn additional income from selling crabs, shrimp and molluscs, while increasing the protein in their diets.

Environmental degradation, combined with heightened exposure and vulnerability of human settlements to risk, contributes to vulnerability to disasters. Almost 2 billion people were affected by disasters in the last decade of the 20th century, 86 per cent of them due to floods and droughts (EM-DAT). Long spells of drought associated with the El Niño Southern Oscillation phenomenon (ENSO) contributed to forest fires in the Amazon Basin, Indonesia and Central America in 1997– 1998.

In Indonesia alone, an estimated 45 600 square kilometres of forest were destroyed. In Central America, the loss of over 15 000 km$^2$ of forests due to wildfires reduced the capacities of natural forests to buffer the impacts of heavy rainfall and hurricanes, and contributed to the devastating impact of Hurricane Mitch in 1998. These impacts spread beyond the tropics, as the large forest fires of California, Spain, Portugal and other Mediterranean countries in 2005 illustrated. Furthermore, coral degradation has negative impacts on coastal communities.

The clustering of climate-related and biological risks will also contribute to impacts on human well-being through events such as heat waves and crop failures. The impact on human health has been addressed in greater detail in the Health section.

**Ecosystems Minimize Risks**

The linkages between biodiversity and livelihood security are complex, and based on the intrinsic relationship between societies and their environment. Policies that can address both the risks and opportunities posed by rapid environmental changes will require a combined focus on ecosystem management, sustainable livelihoods and local risk management.

For example, policies aimed at the improved management of water resources and the non-structural mitigation of weather-related hazards can contribute to the reduction of disaster risks by enhancing landscape restoration, coastal forest management and local conservation and sustainable use initiatives. In coastal ecosystems, restoring mangroves in cyclone-prone areas increases physical protection against storms, creates a reservoir for carbon sequestration and increases livelihood options by generating much-needed income for local communities. Although the evidence base is varied, communities hit by the 2004 tsunami in South Asia reported less damage in areas with healthy mangrove forests than those with few natural sea defences. India and Bangladesh have come to recognize the importance of the Sunderbans mangrove forest in the Gulf of Bengal, not only as a source of livelihoods for fishing communities, but also as an effective mechanism for coastal protection. Viet Nam is also investing in mangrove restoration as a cost-effective means for increased coastal protection. Similar benefits can be derived from coral reefs.

## AGRICULTURE

Links between biodiversity and agriculture Agriculture is defined broadly here to include crops and agroforestry products, livestock and fisheries production. Of some 270 000 known species of higher plants about 10 000–15 000 are edible, and about 7 000 of them are used in agriculture. However, increased globalization threatens to diminish the varieties that are traditionally used in most agricultural systems.

For example, only 14 animal species currently account for 90 per cent of all livestock production, and only 30 crops dominate global agriculture, providing an estimated 90 per cent of the calories consumed by the world's population. Despite its crucial importance in supporting societies, agriculture remains the largest driver of genetic erosion, species loss and conversion of natural habitats around the world.

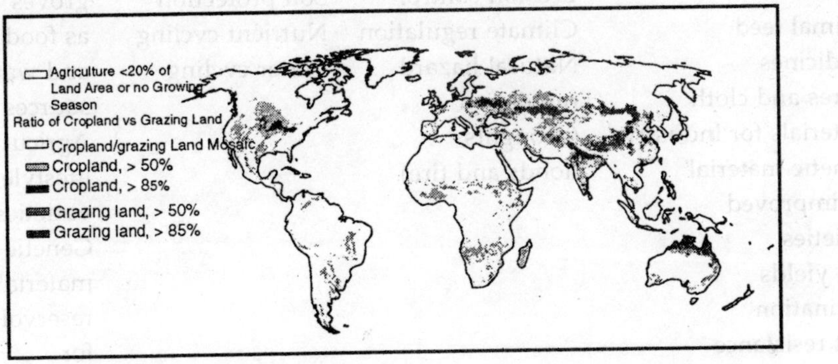


**Fig.** Contemporary Extent of Agricultural Systems

Both cultivated and wild biodiversity provide services necessary for agriculture. Although seldom valued in economic terms, these services play a very significant role in national and regional economies. Different types of agricultural production systems (such as commercial intensive, smallholder, pastoralism and agro-forestry systems) use these services to varying degrees and intensity. For example, the use of nitrogen-fixing legume trees in maize-based systems of Eastern and Southern Africa is helping local farming populations to increase per hectare production of maize without otherwise investing in inorganic fertilizers.

In addition, environmental benefits are gained through carbon sequestration and provision of fuelwood. Habitat conversion is often justified as essential to increasing agricultural production, and trends in agricultural land use over the past 20 years.

Although more than 300 000 km$^2$ of land have been converted to agricultural use in the tropics alone, much of this is of marginal use for agriculture or particular crops. This has led to inefficient use of resources, often resulting in degradation of land and ecosystem services. Some 1.5 billion people, about half of the world's total labour force and nearly onequarter of the global population, are employed in agriculture, or their livelihoods are directly linked to it, and women make up the majority of agricultural workers.

When agriculture on marginal lands is reduced and these lands are appropriately managed, ecosystems can recover, as demonstrated by the expansion of forests in parts of Europe, North America, Japan, China, India, Viet Nam, New Zealand and Latin America.

**Table. Biodiversity Benefits to Agriculture Through Ecosystem Services.**

| Provisioning | Regulating | Supporting | Cultural |
|---|---|---|---|
| Food and nutrients<br>Fuel<br>Animal feed<br>Medicines<br>Fibres and cloth<br>Materials for industry<br>Genetic material for improved varieties and yields<br>Pollination<br>Pest resistance | Pest regulation<br>Erosion control<br>Climate regulation<br>Natural hazard regulation (droughts, floods and fire) | Soil formation<br>Soil protection<br>Nutrient cycling<br>Water cycling | Sacred groves as food and water sources<br>Agricu- lifestyle varieties<br>Genetic material reservoirs for improved varieties and yields<br>Pollinator sanctuaries<br>Erosion control |

Meeting global food needs poses increasing challenges, and will require either intensification or extensification to increase agricultural productivity. Intensified systems tend to be dominated by only a few varieties. This approach is usually associated with higher levels of inputs, including technology, agrochemicals, energy and water use. The latter three, at least, have serious negative impacts on biodiversity.

Extensification relies on lower inputs, and generally on more land being used, often through habitat conversion. In many parts of the world, agricultural extensification involves converting more land for the cultivation of major commodities such as soybeans (Latin America and the Caribbean), oil palm and rubber (Asia and the Pacific), and coffee (Africa, Latin America and Asia), and it is exacerbated by the emergence of new markets for export. In Brazil, for example, the area of land used for growing soybeans (most of which are exported to China) grew from 117 000 $km^2$ in 1994 to 210 000 $km^2$ in 2003.

This was driven by a 52 per cent increase in world consumption of soybeans and soybean products, and these figures continue to rise dramatically. A major agricultural biotechnology innovation during the past two decades is the use of "transgenic" or living modified organisms (LMOs) to provide new attributes in different crops and breeds. The technology is very young, and major investments are being made to enhance its contributions to human well-being and business stability.

Research on LMOs has focused mainly on mitigating the impacts of pests and diseases, and there is evidence of reduced needs for pesticides and herbicides in some crops, such as cotton and maize, through genetic modification. The global production of genetically modified crops (mainly maize, soybean and cotton) was estimated to cover more than 900 000 $km^2$ in 2005.

The use of LMOs is, as for many new technologies, highly controversial, specifically in relation to the uncertain impacts on ecosystems (through escape and naturalization in the landscape), human health and social structures. There are concerns about how its introduction will affect poor people, whose livelihoods depend primarily on traditional low input agricultural practices. Increased research, monitoring and regulation are needed to ensure these negative impacts are avoided as this technology is developed.

The Cartagena Protocol on Biosafety was negotiated and adopted under the CBD to develop a global framework for managing and regulating LMOs. More recently, increasing attention is being given to the existing and potential impacts of climate change on agriculture. Issues include the timing of growth, flowering and maturing of crops, and the impacts of (and on) pollinators, water resources and the distribution of rainfall.

There are also issues of changes in market structures, yields for different crops and strains, and the impacts of extreme weather events on traditional methods and livelihoods. Models show that in some areas, specifically where low temperature is a growth-limiting factor, agricultural productivity may increase with climate change. In other areas, where water and heat are limiting factors, productivity may be severely curtailed. Changes in production practices and loss of diversity in agro-ecosystems can undermine the ecosystem services necessary to sustain agriculture.

For example, pollinator diversity and numbers are affected by habitat fragmentation, agricultural practices, the land-use matrix surrounding agricultural areas and other land-use changes. Although some of the crops that supply a significant proportion of the world's major staples do not require animal pollination (such as rice and maize), the decline of pollinators has long-term consequences for those crop species that serve as crucial sources of micronutrients and minerals (such as fruit trees and vegetables) in many parts of the world.

Genetic erosion, loss of local populations of species, and loss of cultural traditions are often intimately intertwined. While rates of genetic erosion are poorly known, they generally accompany the transition from traditional to commercially developed varieties. In crop and livestock production systems throughout the developing world, genetic erosion reduces smallholder farmer options for mitigating impacts of environmental change and reducing vulnerability, especially in marginal habitats or agricultural systems that are predisposed to extreme weather conditions (such as arid and semi-arid lands of Africa and India).

Implications for agricultural technologies and policy *Methodological and technological innovation* Since the Brundtland Commission report, agricultural research and development has made major advances in integrating conservation and development to mitigate loss of biodiversity, reverse land degradation and foster environmental sustainability. Much remains to be done to create the appropriate enabling environment in many countries, rich and poor alike, especially in eliminating anti-conservation regulations and inappropriate agricultural production subsidies.

A particular area of advancement is the use of innovative agricultural practices to enhance production while conserving native biodiversity. Efforts to foster biodiversity-friendly practices by integrating trees on farms (agro-forestry), conservation agriculture, organic agriculture and integrated pest management are all contributing towards the sustainability of production landscapes. Agroforestry, for example, has emerged as a major opportunity for achieving biodiversity conservation and sustainability in production landscapes, through three major pathways: reducing pressure on natural forests, providing habitat for native plant and animal species and serving as an effective land use in fragmented landscapes.

Integrated land management approaches are also helping to enhance ecosystem resilience through participatory processes that engage and empower farmers, strengthen local institutions and create options for value-added income generation.

These approaches offer significant prospects for restoring degraded lands to enhance habitat connectivity and ecosystem processes. In the tropical forest margins, where slashand- burn farming is a major cause of deforestation, knowledge of land-use dynamics has helped to identify practical options that are profitable for small-scale farmers and at the same time environmentally sustainable.

However, a major challenge to wide-scale implementation of these approaches is the lack of appropriate policy frameworks that align rural and agricultural policies with the protection of biodiversity and ecosystem services. Without such links, the value of integrated natural resource management and eco-agriculture innovations will remain marginal in ensuring the long-term viability of biodiversity.

Very substantial collections of plant genetic resources for food and agriculture are now maintained around the world through the Consultative Group on International Agricultural Research (CGIAR) system. These institutional gene banks are vital for safeguarding germplasm. Farmers have much to contribute at the local level in maintaining the viability of different varieties, such as is being done in an innovative partnership between the International Potato Centre and local communities in Peru, an approach that produces income for the farmers while conserving genetic variability. This also helps maintain local ecological knowledge.

## Policy Options and Governance Mechanisms

Local and community initiatives remain crucial for supporting agricultural approaches that maintain biodiversity. It is challenging to expand these initiatives, since they are based on local differentiation and diversity, rather than homogenization and mass production. The development of recognized standards and certification for production methods can help give producers in these initiatives greater weight and value in the global market. However, little progress has been made overall on institutionalizing a more diverse approach to production systems, and in monitoring its effects.

The techniques that would support reduced pesticide or herbicide use, for example, have yet to be adopted in most countries, and the full value of the ecosystem services provided by ecologically oriented agricultural systems are only very slowly being recognized. Increased research, and the adoption of techniques, such as integrated pest management, can reduce the use of chemicals while providing important biodiversity conservation services. Similarly, remedial measures required to restore productivity to degraded lands are not being implemented on the scale required.

The ecosystem approach can provide a framework for developing practices, such as riparian buffer systems, to both support biodiversity conservation, and assist in water management and purification. National level legislative and policy measures on land tenure and land-use practices will be key to facilitating wide-scale adoption of proven bio-diversity supporting methodologies and technological options in agriculture. The options offer practical solutions that reduce the impacts of agriculture on biodiversity, but need to be considered within a supportive policy framework that encompasses both commercial and small-scale agricultural production landscapes.

At the global scale, ongoing international negotiations are addressing imbalances in markets, subsidies and property rights, all of which have direct links to land use in agriculture. However, there are still major challenges to the conclusion and implementation of the kind of agreements that would generate tangible impacts on biodiversity and agriculture, particularly in the developing world.

### Central America for Biodiversity-friendly Practices

Research into the disappearance of songbirds in the US Midwest is leading to innovations in the production practices and marketing of high-value coffee produced in Central America. Smithsonian Institution researchers found that conversion of forests in Central America for coffee plantations substantially reduced the winter habitats for many migratory birds, reducing their breeding success and their numbers. They worked with coffee producers to test methods of "bird friendly" planting, using intact or minimally-thinned forests for coffee tree planting.

This method of planting produces somewhat fewer coffee beans, but they are of higher quality, and require fewer pesticide and fertilizer inputs. Additionally, the coffee can be marketed as coming from environmentally-friendly sources, potentially bringing in higher prices.

Different certification systems, for example for Bird Friendly® and Shade Grown coffee, show the development and limitations of markets for more sustainably grown crops.

**Multilateral Environmental Biodiversity**

In 1996, parties to the CBD adopted a programme of work on conservation and sustainable use of agricultural biological diversity. In addition, the CBD has established the International Initiative for the Conservation and Sustainable Use of Pollinators, and the International Initiative for the Conservation and Sustainable Use of Soil Biodiversity, both to be implemented in cooperation with FAO and the Global Strategy for Plant Conservation.

Although much remains to be done, global policy processes are helping national governments, particularly in developing countries, to better understand the implications of globalization in agriculture for national policies and development priorities.

The entry into force in June 2004 of the International Treaty on Plant Genetic Resources for Food and Agriculture represents another step in governance of the conservation and use of crop genetic resources, especially for large-scale commercial agriculture. This provides for a multilateral system of exchange for some 30 crops and 40 forage species, and should greatly facilitate use and stimulate the development of effective benefit sharing mechanisms.

# Chapter 9

# Biodiversity and Energy

## LINKS BETWEEN BIODIVERSITY AND ENERGY

Many forms of energy are the result of a service provided by ecosystems, now or laid down in the form of fossil fuels far in the past. Conversely, society's growing requirements for energy are resulting in significant changes in those same ecosystems, both in the search for energy sources, and as a result of energy use patterns. Given that energy is a fundamental requirement for supporting development in all economies, the challenge is to sustainably provide it without driving further loss of biodiversity.

It is necessary to define the trade-offs required, and develop appropriate mitigation and adaptation strategies. Demand for energy is projected to grow at least 53 per cent by 2030. Energy from biomass and waste is projected to supply about 10 per cent of global demand until 2030. However, this assumes that adequate fossil fuels will be available to address the majority of the increase in demand, and some have suggested this may not be realistic. Energy-related carbon dioxide emissions are expected to increase slightly faster than energy use by 2030.

Energy use has impacts at local, national and global levels. Pollution from burning fossil fuels, and the associated effects of acid rain have been a problem for European and North American forests, lakes and soils, although the impacts on biodiversity have not been as significant or widespread as cautioned in the Brundtland Commission report. While emission controls in Europe and North America led to a reversal of acidification trends, there is now a risk of acidification in other areas of the world, particularly Asia.

Use of thermal and nuclear power results in waste disposal problems, as do solar cells, which can result in soil contamination by heavy metals. Desertification in the Sahel and elsewhere in sub-Saharan Africa has been linked in part to fuel demand from biomass. Indirect effects of energy use include both overexploitation of natural resources and greatly facilitated spread of invasive alien species through global trade, both made possible through cheap and easily-available energy for transport.

The impacts noted above are relatively localized and small in comparison to the potential impacts of climate change, which results largely from energy use. As a result of climate change, species ranges and behaviour are changing, with consequences for human well-being, including changing patterns of human disease distribution, and increased opportunities for invasive alien species. Species most likely to be affected include those that already are rare or threatened, migratory species, polar species, genetically impoverished species, peripheral populations and specialized species, including those restricted to alpine areas and islands.

Some amphibian species extinctions have already been linked with climate change, and a recent global study estimated that 15–37 per cent of regional endemic species could be committed to extinction by 2050.

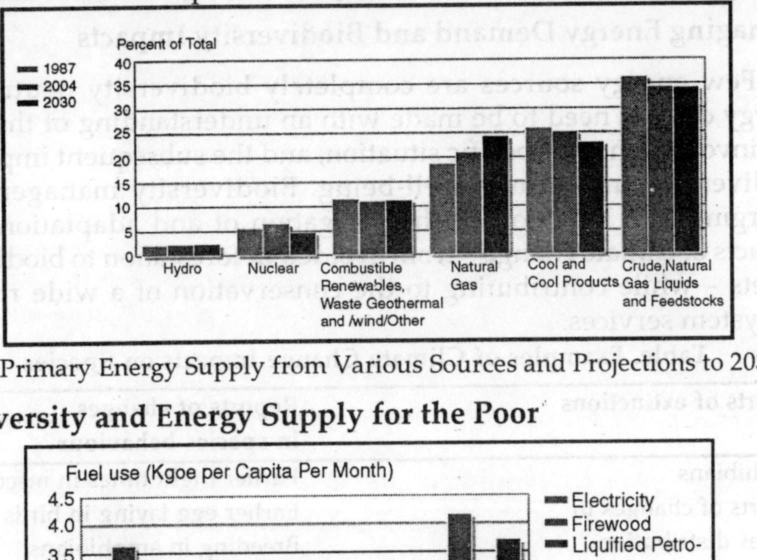


**Fig.** Primary Energy Supply from Various Sources and Projections to 2030.

## Biodiversity and Energy Supply for the Poor

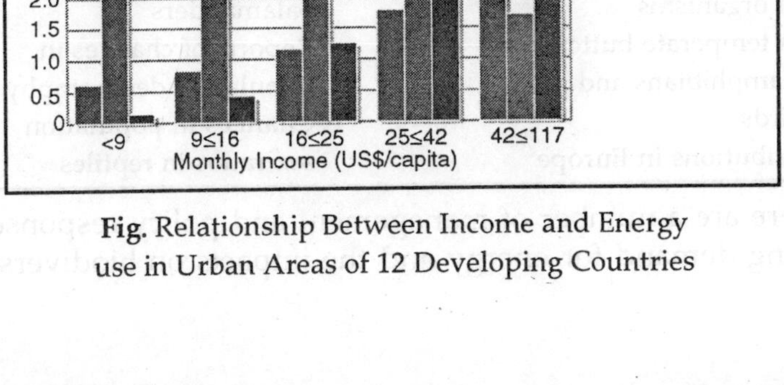


**Fig.** Relationship Between Income and Energy use in Urban Areas of 12 Developing Countries

Biodiversity-based energy sources include both traditional biomass and modern biofuels. Ecosystems provide relatively inexpensive and accessible sources of traditional biomass energy, and therefore have a vital role to play in supporting poor populations.

If these resources are threatened, as is the case in some countries with extreme deforestation, poverty reduction will be an even greater challenge. Use of fuelwood can cause deforestation, but demand for fuelwood can also encourage tree planting, as occurs, for example, in Kenya, Mali and several other developing countries.

Climate change is also having impacts at ecosystem scales. By 2000, 27 per cent of the world's coral reefs had been degraded in part by increased water temperatures, with the largest single cause being the climate-related coral bleaching event of 1998. For some reefs recovery is already being reported. Mediterranean-type ecosystems found in the Mediterranean basin, California, Chile, South Africa and Western Australia are expected to be strongly affected by climate change.

## Managing Energy Demand and Biodiversity Impacts

Few energy sources are completely biodiversity neutral, and energy choices need to be made with an understanding of the trade-offs involved in any specific situation, and the subsequent impacts on biodiversity and human well-being. Biodiversity management is emerging as a key tool for the mitigation of and adaptation to the impacts of climate change – from avoided deforestation to biodiversity offsets – while contributing to the conservation of a wide range of ecosystem services.

**Table. Examples of Climate Change Impacts on Species.**

| Reports of extinctions | Reports of changes in species behaviour |
|---|---|
| Amphibians | Earlier flight times in insects |
| Reports of changes in | Earlier egg laying in birds |
| species distribution | Breeding in amphibians |
| Arctic foxes | Flowering of trees |
| Mountain plants | Ant assemblages |
| Intertidal organisms | Salamanders |
| Northern temperate butterflies | Reports of changes in |
| Tropical amphibians and birds | population demography |
| British birds | Changes in population |
| Tree distributions in Europe | sex ratios in reptiles |

There are a number of management and policy responses to the increasing demand for energy and the impacts on biodiversity. One important response to the rising price of oil is increasing interest in other energy sources. Prime among these are biofuels, with several countries investing significant resources in this field. The world output of biofuels, assuming current practice and policy, is projected to increase almost fivefold, from 20 million tonnes of oil equivalent (Mtoe) in 2005 to 92 Mtoe in 2030.

Biofuels, which are produced on 1 per cent of the world's arable land, support 1 per cent of road transport demand, but that is projected to increase to 4 per cent by 2030, with the biggest increases in United States and Europe. Without significant improvement in productivity of biofuel crops, along with similar progress in food crop agricultural productivity, achieving 100 per cent of transport fuel demand from biofuels is clearly impossible.

In addition, large-scale biofuel production will also create vast areas of biodiversity-poor monocultures, replacing ecosystems such as low-productivity agricultural areas, which are currently of high biodiversity value. Current actions to address the impacts of climate change can be both beneficial and harmful to biodiversity. For example, some carbon sequestration programmes, designed to mitigate impacts of greenhouse gases, can lead to adverse impacts on biodiversity through the establishment of monoculture forestry on areas of otherwise high biodiversity value.

Avoiding deforestation, primarily through forest conservation projects, is an adaptation strategy that may be beneficial, with multiple benefits for climate change mitigation, forest biodiversity conservation, reducing desertification and enhancing livelihoods. It must be recognized that some "leakage" in the form of emissions resulting from those conservation efforts can occur. Climate change will also affect current biodiversity conservation strategies. For example, shifts from one climate zone to another could occur in about half of the world's protected areas, with the effects more pronounced in those at higher latitudes and altitudes.

Some protected area boundaries will need to be flexible if they are to continue to achieve their conservation goals. The impacts of energy production and use on biodiversity have been addressed as a byproduct of several policy responses in the past few decades. Examples include Germany's effort to reduce subsidies in the energy and transport sectors, promoting increases in the proportion of organic farming and reducing nitrogen use in agriculture. However, responses have not been comprehensive, coordinated or universal. Commitments, including shared plans of action, have been made in various fora, but implementation has proved to be extremely challenging, due both to problems of securing required finance and lack of political will or vision.

**Table. Top Biofuel Producers in 2005 (million litres)**

| | |
|---|---|
| **Biodiesel** | |
| Germany | 1 920 |
| France | 511 |
| United States | 290 |
| Italy | 270 |
| Austria | 83 |
| **Bio-ethanol** | |
| Brazil | 16 500 |
| United States | 16 230 |
| China | 2 000 |
| European Union | 950 |
| India | 300 |

**Table. Energy Sources and their Impacts on Biodiversity**

There are also attempts to address this issue through impact management within the private sector, and especially in the energy industry.

The private sector is increasingly accepting its responsibilities as a steward of the environment. It is collaborating with non-governmental organizations, through fora such as the Energy and Biodiversity Initiative, to better understand impacts and possible mitigation and adaptation strategies that make business sense.

Beyond legislation and regulation, the use of payments for ecosystem services, as exemplified by the emerging carbon market, represents an innovative though somewhat controversial approach to addressing the impacts of energy use on the environment. The *State of the Carbon Market 2006*, which covers the period from 1 January, 2005 to 31 March, 2006, records a burgeoning global carbon market, worth over US$10 billion in 2005, 10 times the value of the previous year, and more than the value (US$7.1 billion) of the entire US wheat crop in 2005.

Ensuring access to energy while maintaining biodiversity and vital ecosystem services will require an integrated multi-sectoral approach that includes:

- An ecosystem approach to management of biodiversity and natural resources that ensures inclusion of lessons learned in ongoing management of natural resources affected by energy production and use;
- A major shift in environmental governance to incorporate

| Energy source | Impacts on biodiversity | Subsequent impact on human well-being |
|---|---|---|
| *Fossil fuels*<br>Crude oil<br>Coal<br>Natural gas | Global climate change and associated disturbances, particularly when coupled with human population growth and accelerating rates of resource use, will bring losses in biological diversity.<br>Air pollution (including acid rain) has led to damage to forests in southern China amounting to US$14 billion/year. Losses from air pollution impacts on agriculture are also substantial, amounting to US$4.7 billion in Germany, US$2.7 billion in Poland and US$1.5 billion in Sweden (Myers and Kent 2001).<br>The direct impact of oil spills on aquatic and marine ecosystems are widely reported. The most infamous case is the Exxon Valdez, which ran aground in 1989, spilling 37 000 tonnes of crude oil into Alaska's Prince William Sound (ITOPF 2006).<br>Impacts also come through the development of oil fields and their associated infrastructure, and human activities in remote areas that are valuable for conserving biodiversity (such as Alaska's Arctic National Wildlife Refuge that may be threatened by proposed oil development). | Changes in distribution of and loss of natural resources that support livelihoods.<br>Respiratory disease due to poor air quality. |

policies and incentives promoting energy production and use that mainstreams action to address biodiversity concerns, especially with respect to climate change;

- Increasing partnership with the private sector, including extractive industries and the financial sector, to promote energy programmes that internalize the full costs on biodiversity and livelihoods.

| | | |
|---|---|---|
| *Biomass* Combustibles, renewables and waste | Decreased amount of land available for food crops or other needs due to greatly expanded use of land to produce biofuels, such as sugar cane or fast-growing trees, resulting in possible natural habitat conversion to agriculture, and intensification of formerly extensively developed or fallow land.<br>Can contribute chemical pollutants into the atmosphere that affect biodiversity (Pimentel and others 1994).<br>Burning crop residues as a fuel also removes essential soil nutrients, reducing soil organic matter and the water-holding capacity of the soil.<br>Intensively managing a biofuel plantation may require additional inputs of fossil fuel for machinery, fertilizers and pesticides, with subsequent fossil fuel related impacts.<br>Monoculture of biomass fuel plants can increase soil and water pollution from fertilizer and pesticide use, soil erosion and water run-off, with subsequent loss of biodiversity. | Cardiovascular and respiratory disease from reduced indoor air quality, due to woodburning stoves, especially among poor women and children.<br>Decreased food availability. |

## HEALTH

Biodiversity change affects human health Although there is limited understanding of the consequences of many specific changes in biodiversity for health and the incidence of disease in people and other species, the conceptual links between broader environmental changes and human health are well understood.

Emerging diseases resulting from the destruction and fragmentation of tropical forests and other ecosystems, wildlife-human disease linkages (for example, Lyme disease, West Nile virus and avian influenza), the many known and as yet undiscovered pharmaceutical products found in nature, the contribution of ecosystem services to human health and the increasing recognition of the impacts of endocrine disrupters on both animal and human health, all underline

| | | |
|---|---|---|
| *Nuclear energy* | Water used to cool reactors is released to environment at significantly above ambient temperatures, and accentuates ecological impacts of climatic extremes, such as heat waves, on riverine fauna. Produces relatively small amounts of greenhouse gases during construction. Because of the potential risks posed by nuclear energy, some nuclear plants are surrounded by protected areas. For example, the Hanford Site occupies 145 000 ha in southeastern Washington State. It encompasses several protected areas and sites of long-term research (Gray and Rickard 1989), and provides an important sanctuary for plant and animal populations. A nuclear accident would have grave implications for people and biodiversity. | Health impacts of ionising radiation include deaths and diseases due to genetic damage (including cancers and reproductive abnormalities). |
| *Hydroelectricity* | Building large dams leads to loss of forests, wildlife habitat and species populations, disruption of natural river cycles, and the degradation of upstream catchment areas due to inundation of the reservoir area (WCD 2000). Dam reservoirs also emit greenhouse gases due to the rotting of vegetation and carbon inflows from the basin. On the positive side, some dam reservoirs provide productive fringing wetland ecosystems with fish and waterfowl habitat opportunities. | Building large dams can result in displacement of people. Alterations in availability of freshwater resources (both improved and declining, depending on the situation) for human use. |

the links between biodiversity and human health. About 1 billion people live a subsistence lifestyle, and loss of ecosystem productivity (for example through loss of soil fertility, drought or overfishing) can rapidly lead to malnutrition, stunted childhood growth and development, and increased susceptibility to other diseases. There is

| | | |
|---|---|---|
| *Alternative energy sources* Geothermal Solar, wind, tidal and wave | Ecosystem disruption in terms of desiccation, habitat losses at large wind farm sites and undersea noise pollution. Tidal power plants may disrupt migratory patterns of fish, reduce feeding areas for waterfowl, disrupt flows of suspended sediments and result in various other changes at the ecosystem level. Large photovoltaic farms compete for land with agriculture, forestry and protected areas. Use of toxic chemicals in the manufacture of solar energy cells presents a problem both during use and disposal (Pimentel and others 1994). Disposal of water and wastewater from geothermal plants may cause significant pollution of surface waters and groundwater supplies. Rotors for wind and tidal power can cause some mortality for migratory species, both terrestrial and marine (Dolman and others 2002). Strong visual impact of wind farms. | Decreased species populations to provide basic materials of life. Toxins released to the environment may cause public health problems. Decreased economic value of lands near wind farms, due to strong visual impacts. |

a profound global nutritional imbalance, with a billion overnourished (mainly rich) people and a similar number undernourished (mainly poor people). Historically, this imbalance has been driven primarily by social and economic factors, but ecological factors will probably play an increasingly important role in the future.

Some 70 per cent of infectious diseases originate in animals, and conservation issues are central to their epidemiology. Increased risks of infectious disease spread and crossover can result from land-use changes, many forms of intensive animal production, invasive alien species and the international wildlife trade. Climate change is expanding the range and activity level of disease vectors, particularly insect-borne vectors. The recent international scares over Sudden Acute Respiratory Syndrome (SARS) and avian influenza have brought a dramatic new dimension to the global health debate.

Along with biodiversity changes, there are a number of other factors that are increasing the exposure to and risk of disease. An increasing human population provides an increased number of hosts for disease agents; climate change raises temperatures, altering the wider distribution of disease vectors, such as mosquitoes; drug resistance to conventional treatments is increasing; and continuing poverty and malnutrition make many people more susceptible to disease.

Recent experiences with West Nile virus, hantavirus, avian influenza and tuberculosis provide evidence that disease causing micro-organisms are rapidly adapting to changing circumstances, and emerging or increasing rates of infectious diseases are the result.

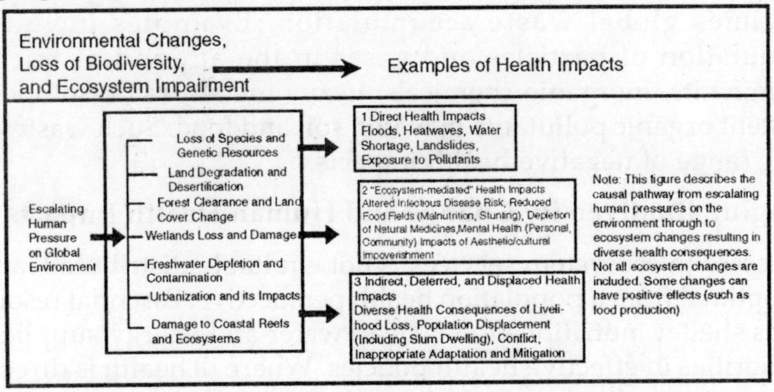


**Fig.** Harmful Effects of Ecosystem Changes on Human Health.

However, changes to ecosystems and their services, especially freshwater sources, food-producing systems and climatic stability, have been responsible for significant adverse impacts on human health in the past 20 years, predominantly in poor countries. Wealthy communities are often able to avoid the effects of local ecosystem degradation by migration, substitution or by appropriation of resources from less-affected regions. Biodiversity is also the source for many cures.

In 2002–2003, 80 per cent of new chemicals introduced globally as drugs could be traced to or were inspired by natural products. Profits from such developments can be enormous. For example, a compound derived from a sea sponge to treat herpes was estimated to be worth US$50–100 million annually, and estimates of the value of anti-cancer agents from marine organisms are up to US$1 billion a year. Traditional medicines, mainly derived from plants, are a mainstay of primary health care for a significant proportion of the population in developing countries.

It is speculated that some 80 per cent of people in developing countries rely on traditional medicines, mostly derived from plants, and more than half of the most frequently prescribed drugs in developed countries derive from natural sources. Loss of biodiversity may decrease our options for new treatments in the future. WHO has identified 20 000 species of medicinal plants for screening, and there are many more species whose medicinal values are only just being discovered, or may prove important in the future.

The value of the global herbal medicine market was estimated at roughly US$43 billion in 2001. The capacity of ecosystems to remove wastes from the environment is being degraded, due to both increased loading of wastes and degradation of ecosystems, leading to local and sometimes global waste accumulation. Examples include the accumulation of particles and gases in the air, and of microbial contaminants, inorganic chemicals, heavy metals, radioisotopes and persistent organic pollutants in water, soil and food. Such wastes have a wide range of negative health impacts.

## Managing Biodiversity Change and Human Health Impacts

Access to ecosystem services is not equitably distributed, and far from optimal from a population health perspective. Essential resources, such as shelter, nutritious food, clean water and energy supplies, are top priorities in effective health policies. Where ill health is directly or indirectly a result of excessive consumption of ecosystem services, substantial reductions in consumption would have major health benefits, and simultaneously reduce pressure on ecosystems.

For example, in rich countries, where overconsumption is causing increasing health impacts, the reduced consumption of animal products and refined carbohydrates would have significant benefits for both human health and for ecosystems globally. Integration of national agricultural and food security policies with the economic, social and environmental goals of sustainable development could be achieved, in part, by ensuring that the environmental and social costs of production and consumption are reflected more fully in the price of food and water.

Responses that mitigate the impacts of ecosystem changes on human health often involve policies and actions outside the health sector. Action to mitigate impacts of climate change will require cooperation across multiple sectors. However, the health sector bears responsibility for communicating the health impacts of ecosystem changes, and of effective and innovative interventions. Where there are tradeoffs, such as between mitigation of negative health impacts and economic growth in other sectors, it is important that the health consequences are well understood, so that they can be included when setting priorities and determining trade-offs.

## CULTURE

Interactions between biodiversity and culture Over the past two decades, there has been growing recognition of the relevance of culture and cultural diversity for the conservation of biodiversity and for sustainable development, as made explicit during the WSSD in 2002. In each society, culture is influenced by locally specific relationships between people and the environment, resulting in varied values, knowledge and practices related to biodiversity.

Cultural knowledge and practices have often contributed specific strategies for the sustainable use and management of biodiversity. The diversity of cultures that have developed globally provides a vast array of responses to different ecosystems, and to variation and change in environmental conditions within them. This cultural diversity forms an essential part of the global pool of resources available to address the conservation of biodiversity. However, cultural diversity is being rapidly lost, in parallel to biological diversity, and largely in response to the same drivers.

Taking linguistic diversity as an indicator of cultural diversity, over 50 per cent of the world's 6 000 languages are currently endangered, and it has been speculated that up to 90 per cent of existing languages may not survive beyond 2100. With loss of languages comes the loss of cultural values, knowledge, innovations and practices, including those related to biodiversity. In addition to the importance of culture for the conservation and sustainable use of biodiversity, human societies everywhere are themselves dependent on biodiversity for their livelihoods, as well as for cultural identity, spirituality, inspiration, aesthetic enjoyment and recreation.

Loss of biodiversity thus affects both material and non-material human well-being. Although societies in industrialized countries may be further removed from the immediate impacts of biodiversity loss, they are nevertheless adversely affected by loss or decline in ecosystem services. Certain categories of people are especially vulnerable to drastic environmental and social change. They include the poor, women, children and youth, rural communities, and indigenous and tribal peoples. The latter constitute the majority of the world's cultural diversity.

Correlations have been identified between the respective geographic distributions of cultural and biological globally and regionally. Figure highlights this, showing the worldwide distributions of plant diversity and linguistic diversity. Areas of high biodiversity tend to be areas of a higher concentration of distinct cultures. Meso-America, the Andes, Western Africa, the Himalayas, and South Asia and the Pacific, in particular, present this pattern of high "biocultural" diversity. This pattern is supported by research that combines indicators of cultural diversity with indicators of biodiversity into a global biocultural diversity index.

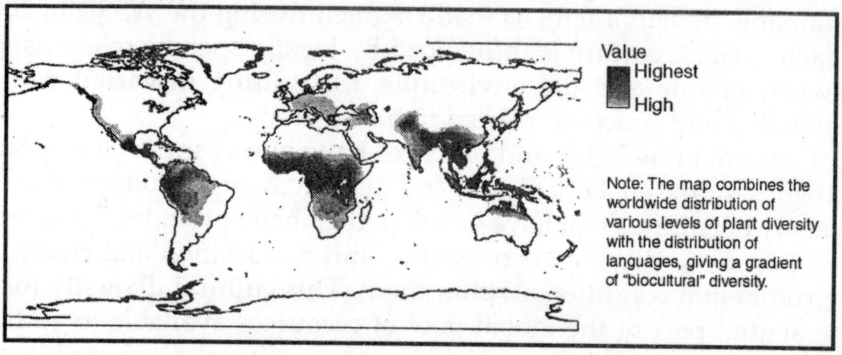


**Fig.** "Biocultural" Diversity in the World

While correlations are evident at the global level, the identification of any causal links between biodiversity and cultural diversity requires research at the local level. Empirical evidence that supports the interrelationships between cultures and biodiversity includes:

- Anthropogenic creation and maintenance of biodiverse landscapes through traditional lowimpact resource management practices;
- Large contribution of traditional farmers to the global stock of plant crop varieties and animal breeds;
- Customary beliefs and behaviours that contribute directly or indirectly to biodiversity conservation, such as sustainable resource extraction techniques, sacred groves, ritual regulation of resource harvests and buffer zone maintenance;
- Dependence of socio-cultural integrity and survival of local communities on access to and tenure of traditional territories, habitats and resources, which also importantly affect food security.

These findings point to significant ecological and societal implications of the increasing threats to the world's cultural diversity. Global social and economic change, is driving the loss of biodiversity, and disrupting local ways of life by promoting cultural assimilation and homogenization.

Cultural change, such as loss of cultural and spiritual values, languages, and traditional knowledge and practices, is a driver that can cause increasing pressures on biodiversity, including over-harvesting, widespread land-use conversion, overuse of fertilizers, reliance on monocultures that replace wild foods and traditional cultivars, and the increase and spread of invasive alien species that displace native species.

In turn, these pressures impact human well-being. The disruption of cultural integrity also impedes the attainment of the Millennium Development Goals (MDGs).

## IMPACTS OF LOSS OF CULTURAL DIVERSITY

**Table. Impacts of loss of cultural diversity**

| Impact on vulnerable groups dependent on local resources | Relevance to MDGs |
|---|---|
| Local food insecurity due to reduction of traditional varieties of crops and access to wild foods (IUCN 1997)<br>Devaluation of gender-specific knowledge of biodiversity, especially women's knowledge of medicines and food sources (Sowerwine 2004)<br>Loss of traditional and local knowledge, practices and language relevant to conservation and sustainable use of biodiversity (Zent and Lopez-Zent 2004) | Goal 1<br>Eradicate extreme poverty and hunger<br>Goal 3<br>Promote gender equality and empower women<br>Goal 7<br>Ensure environmental sustainability |

**Fig.** The Spread of Invasive Alien Species such as the water Hyacinth can have Adverse Impacts on Biodiversity.

**Managing Biological and Cultural Diversity**

The growing recognition over the past two decades of the importance of culture and cultural diversity to the environment and human well-being has led to significant developments in terms of policy and other responses relevant to sustainable development and biodiversity conservation at international, national and local levels.

The policies and activities of UNEP, UNESCO, IUCN and the CBD now include a focus on the interrelationships between biodiversity and cultural diversity, and the indicators for measuring progress towards meeting the CBD's 2010 target include a focus on trends in cultural diversity. In 2006, the UN Human Rights Council adopted the UN Declaration on the Rights of Indigenous Peoples, recognizing that "respect for indigenous knowledge, cultures and traditional practices contributes to sustainable and equitable development and proper management of the environment."

National policies have also taken the initiative to strengthen the links between biodiversity and cultures in accord with the CBD. For example, the Biological Diversity Act of India stipulates that central government shall endeavour to respect and protect the knowledge of local people relating to biodiversity. In doing so, the act provides that forests protected as sacred groves in the context of local communities' belief systems may be recognized as heritage sites.

In Panama, legal recognition has been given in the form of sovereignty to the seven major groups of indigenous peoples in that country. Panama was the first government in Latin America to recognize this class of rights for its indigenous populations, and 22 per cent of the national territory is now designated as sovereign indigenous reserves. Effective biodiversity conservation, particularly that outside of protected areas, relies on integrating local participation, knowledge and values in land-use planning, for example in the co-management of forests, watersheds, wetlands, coastal areas, agricultural lands and rangelands, fisheries, and migratory bird habitats.

Successful comanagement often involves partnerships between local communities and governments, international and local organizations and the private sector, including ecotourism ventures. Incorporating local and traditional knowledge in policy decisions and on-the-ground action calls for mainstreaming the links between biodiversity and culture into social and sectoral plans and policies.

This approach involves developing and strengthening institutions at all scales, so that local knowledge for the conservation and sustainable use of biodiversity can be successfully transferred to landscape and national scales. It also involves strengthening the retention of traditional knowledge through education, conservation of languages and support for passing on knowledge between generations.

An integrative approach to biodiversity conservation for sustainable development takes into account the importance of maintaining the diversity of culturally based knowledge, practices, beliefs and languages that have contributed to the conservation and sustainable use of local biodiversity.

The adoption of this integrative approach in international and national policy directives and on-the-ground interventions signals positive change.

Further recognition of impacts on the most vulnerable societies and social categories of people, and efforts to strengthen the contribution of local and traditional ecological knowledge to policy recommendations, will assist in the maintenance of sustainable relationships between people and biodiversity.

## CHALLENGES AND OPPORTUNITIES

### CHALLENGES

#### Undervaluation of Biodiversity

Biodiversity loss continues because the values of biodiversity are insufficiently recognized by political and market systems. In part this is due to the costs of biodiversity loss not being borne solely by those responsible for its loss.

An added complexity is that the global nature of many biodiversity values results in the impact of biodiversity loss being felt far beyond national boundaries. Losses of biodiversity, such as the erosion of genetic variability in a population, are often slow or gradual, and are often not seen or fully recognized until it is too late.

The dramatic and immediate problems typically receive greater policy attention and budgetary support, so funding is often more available for charismatic megafauna, such as tigers or elephants, than for the wider, yet less celebrated variety of biodiversity that forms key components of the planet's infrastructure, and makes the most substantial contribution to delivering the wide range of ecosystem services from which people benefit. Many of the attempts to calculate the values of biodiversity consider transaction values of the individual components of biodiversity, the price paid for particular goods-and-services.

Although this incorporates some of biodiversity's values, it consistently undervalues many ecosystem functions that are essential for the delivery of ecosystem services. In addition, some elements of biodiversity are irreplaceable when lost, for example through species extinction, or gene loss. Economic valuation and new market mechanisms need to be part of a larger policy toolbox, to take account of such irreversible changes to biodiversity, and although more complete economic valuation is necessary to help create important incentives and opportunities for conservation, it will be insufficient to fully conserve biodiversity for future generations.

Traditional conservation programmes, focused on protecting components of biodiversity from exploitation and other drivers, will remain an important policy tool to protect the irreplaceable and many of the intangible values of biodiversity.

Society can only develop without further loss of biodiversity if market and policy failures are corrected, including perverse production subsidies, undervaluation of biological resources, failure to internalize environmental costs into prices, and failure to recognize global values at the local level.

Most policy sectors have impacts on biodiversity, and biodiversity change has significant implications for those sectors.

However, biodiversity concerns are rarely given sufficient standing when industrial, health, agricultural, development or security policies are developed. Although any society or economy that continues to deplete biodiversity is, by definition, unsustainable, mainstreaming biodiversity concerns effectively into broader policy making so that all policy supports environmental sustainability remains a key challenge. Reducing the rate of biodiversity loss will require multiple and mutually supportive policies of conservation and sustainable use, and the recognition of biodiversity values.

New policies of integrated landscape and watershed management and sustainable use – the ecosystem approach – can be effective in reducing biodiversity loss. In recent years, legal structures such as "biodiversity easements" and "payments for biodiversity services" have been developed to use market mechanisms to provide additional financial resources, and new markets for biodiversity-friendly products are developing new options for producers. These present new opportunities to recognize and mainstream the value of biodiversity, and can address many of the drivers of biodiversity loss.

With a supportive policy framework, such changes will initiate market and behavioral corrections that will move society towards increased sustainability. Although they only make up a small fraction of total market share, organic and sustainable produced agricultural products, such as "bird-friendly" coffee and cocoa, are clear examples of this. However, each of these attempts also has to be cost-effective in the local or global marketplace, and comply with other obligations, such as international trade rules, which often remain perversely disconnected from environmental needs and policies.

**Ineffective Governance Systems**

Political authority and power often reside far from where decisions that affect biodiversity conservation and sustainable use are taken. This includes disjuncture between and within countries, where different ministries frequently take different approaches to the issue of biodiversity management. Biodiversity concerns are dealt with in numerous international and regional agreements, many of which have come into force in the past 20 years.

In 2004, five of the key global biodiversity-related conventions (CBD, CITES, CMS, Ramsar and the World Heritage Convention) created the Biodiversity Liaison Group to help facilitate a more coordinated approach to policy development and implementation. UNEP has created the Issues-based Modules project, which aims to assist countries and other stakeholders to understand the intersections of obligations coming from the various conventions.

Such actions and projects epitomize the call from the WSSD to shift from policy development to implementation, and provide a start towards an integrated approach to biodiversity management. Biodiversity governance involves multiple stakeholders, including landholders, community and political jurisdictions (local, national and regional), the private sector, specific arrangements such as fisheries management councils, species protection agreements and the global agreements. Most of these suffer from a lack of financial and human capacity to effectively manage biodiversity.

Even very clear policies do not ensure compliance or enforcement, as is evident from the ongoing illegal international trade in species and their parts, in contravention of CITES. The proliferation of authorities has, in many instances, created confusion, dispersed resources, and slowed policy development and implementation. This has led to coordination problems between and within scales: local to national, inter-ministerial, regional and international. In most countries, biodiversity concerns are the responsibility of relatively weak, underfunded and understaffed environmental ministries.

Decisions that severely threaten biodiversity, such as land-use changes and the introduction of potentially invasive species (either by design or accident), are most frequently taken by agriculture, fisheries, commerce or mining ministries. Often this is done without effective consultation with the authorities responsible for the environment, or recognition of the costs of such impacts. Biodiversity governance is in a major period of flux. Historically, biodiversity was largely considered as a common heritage, and a public good.

The late 20$^{th}$ century saw an unprecedented "enclosing" of genetic resources, a shift from considering them common heritage to seeing them as products to be owned in whole or in part. Two components of this recent enclosure movement are the patenting of genes, gene expressions and derived life forms on the one hand, and the fundamental shift to the concept of ownership of genetic resources that arose through the CBD and the FAO International Treaty on Plant Genetic Resources, in terms of national sovereignty over biological diversity.

At the same time, the importance of biodiversity is better recognized, not only as a source of new products, but also as fundamental for the supply of the full range of ecosystem services. In 2002, the CBD adopted the Bonn Guidelines on Access to Genetic Resources and Fair and Equitable Sharing of the Benefits Arising out of their Utilization (ABS), and the WSSD subsequently called for further elaboration of the international regime on access and benefit sharing.

Although the resulting negotiations have since dominated much of the international biodiversity discourse, the "green gold" predicted by the early advocates of the issue within the CBD, and the "gene gold" predicted by the rush to patent genetic information have not materialized. Whether this reflects an early market or overinflated predictions is not clear. However, these ABS discussions are likely to continue to dominate international negotiations, not only on biodiversity, but also on trade and intellectual property, distracting discussion from other fundamental issues of greater importance to the sustainable supply of ecosystem services for development.

Further research and understanding on how to capture and distribute the benefits arising from the use of biodiversity will contribute to these discussions. The CBD has taken a novel and progressive approach to identify a mechanism to respect the breadth of traditional knowledge on the uses for biodiversity. The strengthened voices of indigenous communities have brought forward important and as yet unresolved issues, including tensions between different ways of knowing (western science and community cosmology), valuing (economically based and culturally based) and governing (formalized written and customary law).

Local and indigenous communities, and women within them, have been and will continue to be important stewards of biodiversity, and national systems of land tenure and respect for indigenous communities are intertwined with biodiversity policy making at the local and international levels.

## OPPORTUNITIES

New and evolving concepts of ownership over biodiversity and genetic resources, protection of traditional knowledge, the ecosystem approach, ecosystem services and valuation, have created policy challenges for all of the actors. Governments at all levels, communities and businesses are grappling with how to incorporate environmental, social and cultural concerns more effectively into their decision making processes. In order to achieve sustainable development, biodiversity needs to be mainstreamed into energy, health, security, agricultural, land use, urban planning and development policies.

### Management Interlinkages

At the international level, the biodiversity-related conventions have increased their collaboration, and are attempting to link more closely with economic instruments such as World Intellectual Property Organization and the World Trade Organization. Each of these processes has developed strategies and action plans that need to be implemented nationally, and there is a clear need to find which approaches work best, under which circumstances, and to deliver more effective advice at each level.

### Private Sector Interventions

Some private corporations have started to build biodiversity concerns into their planning and implementation, but many more still need to analyse and minimize the negative impacts of infrastructure development, and operations, such as processing and transportation, on biodiversity. Seemingly good policy may mask environmental degradation elsewhere, such as the movement of polluting industries to, or sourcing of wood products from, less-regulated areas.

Codes of conduct, certification schemes, transparency through triple-bottom line accounting and international regulatory standards are key policy options for creating incentives and level playing fields that will minimize these cost-shifting behaviours. Regional organizations, such as the European Community, NAFTA and SADC, play important roles in creating such level playing fields, and collaboration across sectors within government is also required. Interagency coordination is needed to bring coherence to international negotiations, and to bring biodiversity concerns into national policy development.

### *Market Mechanisms*

Appropriately recognizing the multiple values of biodiversity in national policies is likely to require new regulatory and market mechanisms, such as:

- Better valuation and the creation of markets for ecosystem services;
- More widespread certification systems;
- payment programmes to increase incentives for conservation and protection of biodiversity and ecosystems;
- New policies providing tax incentives for low biodiversity impact operations;
- Reducing and eliminating perverse incentives for biodiversity loss;
- Developing conservation easements;
- Mechanisms for upstream-downstream transfers.

### *Pro-poor Policies*

Implementing policies that benefit the poorest in society will be challenging, but necessary. Raising the profile and representation of direct biodiversity users and stewards, especially smallholders, will be key in developing effective implementation mechanisms.

Recognizing the role women play in protecting, using and understanding biodiversity in many parts of the world can lead to the mutual benefit of empowering communities and ensuring sustainable use of biodiversity. Including all stakeholders in the shaping and testing of policies will be necessary to ensure long-term viability and acceptance of the policy changes. Generalizing and scaling up inclusive projects is a key challenge and opportunity for the international community.

### *Conservation Measures*

Natural disasters in recent years – tsunamis, hurricanes and earthquakes – have highlighted a range of environmental and biodiversity concerns. Preservation and restoration of coastal mangroves, seagrasses, coastal wetlands and reef systems protect shorelines from the power of storms.

Forests regulate water flow, and soil structure and stability. Policies that help protect biodiversity also protect people and infrastructure. Taking the range of biodiversity and environmental concerns into account in land-use planning, and enforcing rules and regulations are key to success.

*New Governance Structures*

The understanding of biodiversity, its role and uses, and the governance structure of enclosure is all in its infancy, as nations and localities are testing options, and finding opportunities and obstacles. Further analysis and assessment of valuation programmes, mainstreaming attempts, and new governance structures are needed to develop best practices and share lessons learned. As more policy tools and mechanisms based on success are developed, new ways will emerge to conserve and use the world's biodiversity.

However, enough is already known to make better decisions on the conservation and wise use of biodiversity. Given the documented rate of habitat conversion and degradation, and declines in populations and genetic resources, much more action is needed immediately to conserve biodiversity so that future generations will have the full range of opportunities to benefit from its use.

## Information Gaps and Research Needs

Due to the complexity of the concept of biodiversity there is no simple list of information gaps that, if filled, would answer the majority of the questions this chapter has raised. However, each level has some significant information needs, and addressing these would provide multiple benefits:

*What Exists on Earth and Where?*

These fundamental questions of description and biogeography underpin all biodiversity and ecosystem research. The discovering, naming, describing and ordering of the different species on Earth is a science called taxonomy. It is needed, for example, to identify invasive species, differentiate between different disease vectors and reservoirs, and identify likely candidates for new medicines and other useful chemicals and enzymes.

However, the majority of the world's species have not been identified, and some key groups, such as invertebrates and micro-organisms, are especially poorly understood. The CBD has created the Global Taxonomy Initiative (GTI) to try to overcome this impediment, and the Global Biodiversity Information Facility (GBIF) has been created to pull together the disparate data from taxonomic institutions around the world for integrated use, leveraging each country's investment for the common good. However, greater financial and collaborative support from governments and civil society is needed for these efforts.

### *How do Biological Resources Function?*

From the genetic level through to research on how different organisms move, and process food, water, salt and other inputs (including pollutants), there is an increasing understanding of the range of processes that nature has developed, and that can be used to move towards a more sustainable development path. Examples include:

- The increasing understanding of the genetics of key agricultural organisms, such as rice and potato, which should contribute to the development of more hardy and prolific strains;
- The study of the ability of different classes of microbes to perform a range of functions, from breaking down pollutants to isolating and purifying metals;
- The identification of processes that will allow people to most effectively develop technologies, such as biofuels, without further damaging the environment or harming food security.

Considerable resources are going into this range of research, frequently driven by specific economic interests, but the work is often hampered by a lack of taxonomic and biogeographic understanding.

### *How Does the System Interact?*

The multitude of questions about ecology range from the very local (how do soil microbes support plant growth) to global (how do forest and ocean organisms sequester carbon and regulate climate systems). Answering these questions, and understanding the dynamics within them frequently takes many years' research with repeated observations. In many areas increased research is needed, for example, on:

- the impacts of fragmentation on biodiversity structure and functioning, resiliency of ecosystems to change (such as from climate change and human interventions);
- the role of biodiversity in mitigating and responding to climate change;
- the role of restoration ecology in remediating changed and degraded lands;
- reservoirs and vectors of pathogens and zoonotic diseases.

New mechanisms are also needed to bring together the vast research results in a way to use the data for new modelling and research questions.

### How do People use and Understand Biodiversity?

The vast array of different cultures, and the associated range of knowledge about biodiversity, contributes key understanding for conservation and sustainable use of biodiversity. Many new governance structures and techniques are being developed, and these need to be understood more clearly if their effectiveness and synergies are to be maximized, and the spread of perverse incentives is to be avoided.

There is a need for increased capacity building, to convert knowledge into practice in many parts of the world. Increased understanding of how people relate to biodiversity, and how to move towards greater stewardship of biodiversity may be the biggest question the world still must answer.

### How can Biodiversity be Valued?

Substantial research on internalizing the values of biodiversity, and the adoption of new indices of global and national wealth based on functioning ecosystems are required, including clear and consistent rules and processes that cross economic and political jurisdictions, such as are emerging in areas of forest and organic certification.

## Chapter 10

# Bioremediation

The quality of life on Earth is linked inextricably to the overall quality of the environment. In early times, we believed that we had an unlimited abundance of land and resources; today, however, the resources in the world show, in greater or lesser degree, our carelessness and negligence in using them. The problems associated with contaminated sites now assume increasing prominence in many countries. Contaminated lands generally result from past industrial activities when awareness of the health and environmental effects connected with the production, use, and disposal of hazardous substances were less well recognized than today. The problem is worldwide, and the estimated number of contaminated sites is significant.

It is now widely recognized that contaminated land is a potential threat to human health, and its continual discovery over recent years has led to international efforts to remedy many of these sites, either as a response to the risk of adverse health or environmental effects caused by contamination or to enable the site to be redeveloped for use. The conventional techniques used for remediation have been to dig up contaminated soil and remove it to a landfill, or to cap and contain the contaminated areas of a site.

The methods have some drawbacks. The first method simply moves the contamination elsewhere and may create significant risks in the excavation, handling, and transport of hazardous material. Additionally, it is very difficult and increasingly expensive to find new landfill sites for the final disposal of the material. The cap and contain method is only an interim solution since the contamination remains on site, requiring monitoring and maintenance of the isolation barriers long into the future, with all the associated costs and potential liability. A better approach than these traditional methods is to completely destroy the pollutants if possible, or at least to transform them to innocuous substances. Some technologies that have been used are high-temperature incineration and various types of chemical decomposition (e.g., base-catalyzed dechlorination, UV oxidation). They can be very effective at reducing levels of a range of contaminants, but have several drawbacks, principally their technological complexity, the cost for small-scale application, and the lack of public acceptance, especially for incineration that may increase the exposure to contaminants for both the workers at the site and nearby residents. Bioremediation is an option that offers the possibility to destroy or render harmless various contaminants using natural biological activity. As such, it uses relatively low-cost, low-technology techniques, which generally have a high public acceptance and can often be carried out on site.

It will not always be suitable, however, as the range of contaminants on which it is effective is limited, the time scales involved are relatively long, and the residual contaminant levels achievable may not always be appropriate. Although the methodologies employed are not technically complex, considerable experience and expertise may be required to design and implement a successful bioremediation programme, due to the need to thoroughly assess a site for suitability and to optimize conditions to achieve a satisfactory result. Because bioremediation seems to be a good alternative to conventional clean-up technologies research in this field, especially in the United States, rapidly increasing. Bioremediation has been used at a number of sites worldwide, including Europe, with varying degrees of success.

Techniques are improving as greater knowledge and experience are gained, and there is no doubt that bioremediation has great potential for dealing with certain types of site contamination. Unfortunately, the principles, techniques, advantages, and disadvantages of bioremediation are not widely known or understood, especially among those who will have to deal directly with bioremediation proposals, such as site owners and regulators.

Here, we intended to assist by providing a straightforward, pragmatic view of the processes involved in bioremediation, the pros and cons of the technique, and the issues to be considered when dealing with a proposal for bioremediation. Some tests make an exhaustive examination of the literature of bioremediation of organic and inorganic pollutants, and another test takes a look at pertinent field application case histories.

## PRINCIPLES OF BIOREMEDIATION

Environmental biotechnology is not a new field; composting and wastewater treatments are familiar examples of old environmental biotechnologies. However, recent studies in molecular biology and ecology offer opportunities for more efficient biological processes.

Notable accomplishments of these studies include the clean-up of polluted water and land areas. Bioremediation is defined as the process whereby organic wastes are biologically degraded under controlled conditions to an innocuous state, or to levels below concentration limits established by regulatory authorities.

By definition, bioremediation is the use of living organisms, primarily microorganisms, to degrade the environmental contaminants into less toxic forms. It uses naturally occurring bacteria and fungi or plants to degrade or detoxify substances hazardous to human health and/or the environment. The microorganisms may be indigenous to a contaminated area or they may be isolated from elsewhere and brought to the contaminated site. Contaminant compounds are transformed by living organisms through reactions that take place as a part of their metabolic processes. Biodegradation of a compound is often a result of the actions of multiple organisms.

When microorganisms are imported to a contaminated site to enhance degradation we have a process known as bioaugmentation. For bioremediation to be effective, microorganisms must enzymatically attack the pollutants and convert them to harmless products. As bioremediation can be effective only where environmental conditions permit microbial growth and activity, its application often involves the manipulation of environmental parameters to allow microbial growth and degradation to proceed at a faster rate. Like other technologies, bioremediation has its limitations. Some contaminants, such as chlorinated organic or high aromatic hydrocarbons, are resistant to microbial attack.

They are degraded either slowly or not at all, hence it is not easy to predict the rates of clean-up for a bioremediation exercise; there are no rules to predict if a contaminant can be degraded. Bioremediation techniques are typically more economical than traditional methods such as incineration, and some pollutants can be treated on site, thus reducing exposure risks for clean-up personnel, or potentially wider exposure as a result of transportation accidents. Since bioremediation is based on natural attenuation the public considers it more acceptable than other technologies. Most bioremediation systems are run under aerobic conditions, but running a system under anaerobic conditions may permit microbial organisms to degrade otherwise recalcitrant molecules.

**Factors of Bioremediation**

The control and optimization of bioremediation processes is a complex system of many factors. These factors include: the existence of a microbial population capable of degrading the pollutants; the availability of contaminants to the microbial population; the environment factors (type of soil, temperature, pH, the presence of oxygen or other electron acceptors, and nutrients).

## Microbial Populations for Bioremediation Processes

Microorganisms can be isolated from almost any environmental conditions. Microbes will adapt and grow at subzero temperatures, as well as extreme heat, desert conditions, in water, with an excess of oxygen, and in anaerobic conditions, with the presence of hazardous compounds or on any waste stream. The main requirements are an energy source and a carbon source. Because of the adaptability of microbes and other biological systems, these can be used to degrade or remediate environmental hazards.

**Table. Some Contaminants Potentially Suitable for Bioremediation.**

| Class of contaminants | Specific examples | Aerobic | Anaerobic | More potential sources |
|---|---|---|---|---|
| Chlorinated solvents | Trichloroethylene | | + | Drycleaners |
| | Perchloroethylene | | | Chemical |
| manufacture | | | | |
| Polychlorinated biphenyls | 4-Chlorobiphenyl | | + | Electrical manufacturing |
| | 4,4-Dichlorobiphenyl | | | Power station |
| | | | | Railway yards |
| Chlorinated phenol | Pentachlorophenol | | + | Timber treatment |
| | | | | Landfills |
| "BTEX" | Benzene | + | + | Oil production and storage |
| | Toluene | | | Gas work sites |
| | Ethylbenzene | | | Airports |
| | Xylene | | | Paint manufacture |
| | | | | Port facilities |
| | | | | Railway yards |
| | | | | Chemical |
| manufacture | | | | |
| Polyaromatic hydrocarbons | Naphthalene | + | | Oil production and storage |
| (PAHs) | Antracene | | | Gas work sites |
| | Fluorene | | | Coke plants |
| | Pyrene | | | Engine works |
| | Benzo(a)pyrene | | | Landfills |
| | | | | Tar production and storage |

| | | | | |
|---|---|---|---|---|
| | | | | Boiler ash dump |
| sites | | | | |
| | | | | Power stations |
| Pesticides | Atrazine | + | + | Agriculture |
| | Carbaryl | | | Timber treatment plants |
| | Carbofuran | | | Pesticide |
| manufacture | | | | |
| | Coumphos | | | Recreational areas |
| | Diazinon | | | Landfills |
| | Glycophosphate | | | |
| | Parathion | | | |
| | Propham | | | |
| | 2,4-D | | | |

We can subdivide these microorganisms into the following groups:

- *Aerobic*: In the presence of oxygen. Examples of aerobic bacteria recognized for their degradative abilities are *Pseudomonas, Alcaligenes, Sphingomonas, Rhodococcus*, and *Mycobacterium*. These microbes have often been reported to degrade pesticides and hydrocarbons, both alkanes and polyaromatic compounds. Many of these bacteria use the contaminant as the sole source of carbon and energy.
- *Anaerobic*: In the absence of oxygen. Anaerobic bacteria are not as frequently used as aerobic bacteria. There is an increasing interest in anaerobic bacteria used for bioremediation of polychlorinated biphenyls (PCBs) in river sediments, dechlorination of the solvent trichloroethylene (TCE), and chloroform.
- *Ligninolytic fungi*: Fungi such as the white rot fungus *Phanaerochaete chrysosporium* have the ability to degrade an extremely diverse range of persistent or toxic environmental pollutants. Common substrates used include straw, saw dust, or corn cobs.
- *Methylotrophs*: Aerobic bacteria that grow utilizing methane for carbon and energy. The initial enzyme in the pathway for aerobic degradation, methane monooxygenase, has a broad substrate range and is active against a wide range of compounds, including the chlorinated aliphatics trichloroethylene and 1,2-dichloroethane. An overview of the microbiological aspects of the application of microorganisms. For degradation it is necessary that bacteria and the contaminants be in contact.

This is not easily achieved, as neither the microbes nor contaminants are uniformly spread in the soil. Some bacteria are mobile and exhibit a chemotactic response, sensing the contaminant and moving toward it.

Other microbes such as fungi grow in a filamentous form toward the contaminant. It is possible to enhance the mobilization of the contaminant utilizing some surfactants such as sodium dodecyl sulphate (SDS).

## ENVIRONMENTAL FACTORS

### Nutrients

Although the microorganisms are present in contaminated soil, they cannot necessarily be there in the numbers required for bioremediation of the site.

Their growth and activity must be stimulated. Biostimulation usually involves the addition of nutrients and oxygen to help indigenous microorganisms.

These nutrients are the basic building blocks of life and allow microbes to create the necessary enzymes to break down the contaminants.

All of them will need nitrogen, phosphorous, and carbon. Carbon is the most basic element of living forms and is needed in greater quantities than other elements. In addition to hydrogen, oxygen, and nitrogen it constitutes about 95% of the weight of cells.

**Table. Composition of a Microbial Cell.**

| Element | Percentage | Element | Percentage |
|---|---|---|---|
| Carbon | 50 | Sodium | 1 |
| Nitrogen | 14 | Calcium | 0,5 |
| Oxygen | 20 | Magnesium | 0,5 |
| Hydrogen | 8 | Chloride | 0,5 |
| Phosphorous | 3 | Iron | 0,2 |
| Sulfur | 1 | All others | 0,3 |
| Potassium | 1 | | |

Phosphorous and sulfur contribute with 70% of the remainders. The nutritional requirement of carbon to nitrogen ratio is 10:1, and carbon to phosphorous is 30:1.

### Environmental Requirements

Optimum environmental conditions for the degradation of contaminants are reported in Table.

**Table. Environmental Conditions Affecting Degradation.**

| Parameters | Condition required microbial activity | Optimum value for an oil for degradation |
|---|---|---|
| Soil moisture | 25–28% of water holding capacity | 30–90% |
| Soil pH | 5.5–8.8 | 6.5–8.0 |
| Oxygen content | Aerobic, minimum air-filled pore space of 10% | 10–40% |
| Nutrient content | N and p for microbial growth | C:N:P = 100:10:1 |
| Temperature (°C) | 15–45 | 20–30 |
| Contaminants | Not too toxic | Hydrocarbon 5–10% of dry weight of soil |
| Heavy metals | Total content 2000 ppm | 700 ppm |
| Type of soil | Low clay or silt content | |

Microbial growth and activity are readily affected by pH, temperature, and moisture. Although microorganisms have been also isolated in extreme conditions, most of them grow optimally over a narrow range, so that it is important to achieve optimal conditions. If the soil has too much acid it is possible to rinse the pH by adding lime. Temperature affects biochemical reactions rates, and the rates of many of them double for each 10 °C rise in temperature. Above a certain temperature, however, the cells die. Plastic covering can be used to enhance solar warming in late spring, summer, and autumn.

Available water is essential for all the living organisms, and irrigation is needed to achieve the optimal moisture level. The amount of available oxygen will determine whether the system is aerobic or anaerobic. Hydrocarbons are readily degraded under aerobic conditions, whereas chlorurate compounds are degraded only in anaerobic ones. To increase the oxygen amount in the soil it is possible to till or sparge air. In some cases, hydrogen peroxide or magnesium peroxide can be introduced in the environment. Soil structure controls the effective delivery of air, water, and nutrients. To improve soil structure, materials such as gypsum or organic matter can be applied. Low soil permeability can impede movement of water, nutrients, and oxygen; hence, soils with low permeability may not be appropriate for *in situ* clean-up techniques.

## BIOREMEDIATION STRATEGIES

Different techniques are employed depending on the degree of saturation and aeration of an area. *In situ* techniques are defined as those that are applied to soil and groundwater at the site with minimal disturbance. *Ex situ* techniques are those that are applied to soil and groundwater at the site which has been removed from the site via excavation (soil) or pumping (water). *Bioaugmentation* techniques involve the addition of microorganisms with the ability to degrade pollutants.

### *In Situ* Bioremediation

These techniques are generally the most desirable options due to lower cost and less disturbance since they provide the treatment in place avoiding excavation and transport of contaminants. *In situ* treatment is limited by the depth of the soil that can be effectively treated. In many soils effective oxygen diffusion for desirable rates of bioremediation extend to a range of only a few centimeters to about 30 cm into the soil, although depths of 60 cm and greater have been effectively treated in some cases. The most important land treatments are:

- *Bioventing* is the most common *in situ* treatment and involves supplying air and nutrients through wells to contaminated soil to stimulate the indigenous bacteria. Bioventing employs low air flow rates and provides only the amount of oxygen necessary for the biodegradation while minimizing volatilization and release of contaminants to the atmosphere. It works for simple hydrocarbons and can be used where the contamination is deep under the surface.
- *In situ biodegradation* involves supplying oxygen and nutrients by circulating aqueous solutions through contaminated soils to stimulate naturally occurring bacteria to degrade organic contaminants. It can be used for soil and groundwater. Generally, this technique includes conditions such as the infiltration of water-containing nutrients and oxygen or other electron acceptors for groundwater treatment.
- *Biosparging*. Biosparging involves the injection of air under pressure below the water table to increase groundwater oxygen concentrations and enhance the rate of biological degradation of contaminants by naturally occurring bacteria. Biosparging increases the mixing in the saturated zone and thereby increases the contact between soil and groundwater. The ease and low cost of installing small-diameter air injection points allows considerable flexibility in the design and construction of the system.

- *Bioaugmentation*. Bioremediation frequently involves the addition of microorganisms indigenous or exogenous to the contaminated sites. Two factors limit the use of added microbial cultures in a land treatment unit:
  - Nonindigenous cultures rarely compete well enough with an indigenous population to develop and sustain useful population levels
  - Most soils with long-term exposure to biodegradable waste have indigenous microorganisms that are effective degrades if the land treatment unit is well managed.

## *Ex Situ* Bioremediation

These techniques involve the excavation or removal of contaminated soil from ground.

- *Landfarming* is a simple technique in which contaminated soil is excavated and spread over a prepared bed and periodically tilled until pollutants are degraded. The goal is to stimulate indigenous biodegradative microorganisms and facilitate their aerobic degradation of contaminants. In general, the practice is limited to the treatment of superficial 10–35 cm of soil. Since landfarming has the potential to reduce monitoring and maintenance costs, as well as clean-up liabilities, it has received much attention as a disposal alternative.
- *Composting* is a technique that involves combining contaminated soil with nonhazardous organic amendants such as manure or agricultural wastes. The presence of these organic materials supports the development of a rich microbial population and elevated temperature characteristic of composting.
- *Biopiles* are a hybrid of landfarming and composting. Essentially, engineered cells are constructed as aerated composted piles. Typically used for treatment of surface contamination with petroleum hydrocarbons they are a refined version of landfarming that tend to control physical losses of the contaminants by leaching and volatilization. Biopiles provide a favorable environment for indigenous aerobic and anaerobic microorganisms.
- *Bioreactors*. Slurry reactors or aqueous reactors are used for *ex situ* treatment of contaminated soil and water pumped up from a contaminated plume. Bioremediation in reactors involves the processing of contaminated solid material (soil, sediment, sludge) or water through an engineered

containment system. A slurry bioreactor may be defined as a containment vessel and apparatus used to create a three-phase (solid, liquid, and gas) mixing condition to increase the bioremediation rate of soil bound and water-soluble pollutants as a water slurry of the contaminated soil and biomass (usually indigenous microorganisms) capable of degrading target contaminants.

**Table. Summary of Bioremediation Strategies.**

| Technology | Examples | Benefits | Limitations | Factors to consider |
|---|---|---|---|---|
| *In situ* | *In situ* bioremediation<br>Biosparging<br>Bioventing<br>Bioaugmentation | Most cost efficient<br>Noninvasive<br>Relatively passive<br>Natural attenuation processes<br>Treats soil and water | Environmental constraints<br>Extended treatment time<br>Monitoring difficulties | Biodegradative abilities of indigenous microorganisms<br>Presence of metals and other inorganics<br>Environmental parameters<br>Biodegradability of pollutants<br>Chemical solubility<br>Geological factors<br>Distribution of pollutants |
| *Ex situ* | Landfarming<br>Composting<br>Biopiles | Cost efficient<br>Low cost<br>Can be done on site | Space requirements<br>Extended treatment time<br>Need to control abiotic loss<br>Mass transfer problem<br>Bioavailability limitation | See above |
| Bioreactors | Slurry reactors<br>Aqueous reactors | Rapid degradation kinetic<br>Optimized environmental parameters<br>Enhances mass transfer<br>Effective use of inoculants and surfactants | Soil requires excavation<br>Relatively high cost capital<br>Relatively high operating cost | See above<br>Bioaugmentation<br>Toxicity of amendments<br>Toxic concentrations of contaminants |

In general, the rate and extent of biodegradation are greater in a bioreactor system than *in situ* or in solid-phase systems because the contained environment is more manageable and hence more controllable and predictable.

Despite the advantages of reactor systems, there are some disadvantages. The contaminated soil requires pre treatment (e.g., excavation) or alternatively the contaminant can be stripped from the soil via soil washing or physical extraction (e.g., vacuum extraction) before being placed in a bioreactor. Table summarizes the advantages and disadvantages of bioremediation.

- Bioremediation is a natural process and is therefore perceived by the public as an acceptable waste treatment process for contaminated material such as soil. Microbes able to degrade the contaminant increase in numbers when the contaminant is present; when the contaminant is degraded, the biodegradative population declines. The residues for the treatment are usually harmless products and include carbon dioxide, water, and cell biomass.
- Theoretically, bioremediation is useful for the complete destruction of a wide variety of contaminants. Many compounds that are legally considered to be hazardous can be transformed to harmless products. This eliminates the chance of future liability associated with treatment and disposal of contaminated material.
- Instead of transferring contaminants from one environmental medium to another, for example, from land to water or air, the complete destruction of target pollutants is possible.
- Bioremediation can often be carried out on site, often without causing a major disruption of normal activities. This also eliminates the need to transport quantities of waste off site and the potential threats to human health and the environment that can arise during transportation.
- Bioremediation can prove less expensive than other technologies that are used for clean-up of hazardous waste.

## Disadvantages of Bioremediation

- Bioremediation is limited to those compounds that are biodegradable. Not all compounds are susceptible to rapid and complete degradation.
- There are some concerns that the products of biodegradation may be more persistent or toxic than the parent compound.
- Biological processes are often highly specific. Important site

factors required for success include the presence of metabolically capable microbial populations, suitable environmental growth conditions, and appropriate levels of nutrients and contaminants.

- It is difficult to extrapolate from bench and pilot-scale studies to full-scale field operations.
- Research is needed to develop and engineer bioremediation technologies that are appropriate for sites with complex mixtures of contaminants that are not evenly dispersed in the environment. Contaminants may be present as solids, liquids, and gases.
- Bioremediation often takes longer than other treatment options, such as excavation and removal of soil or incineration.
- Regulatory uncertainty remains regarding acceptable performance criteria for bioremediation. There is no accepted definition of "clean", evaluating performance of bioremediation is difficult, and there are no acceptable endpoints for bioremediation treatments.

## PHYTOREMEDIATION

Although the application of microbe biotechnology has been successful with petroleum-based constituents, microbial digestion has met limited success for widespread residual organic and metals pollutants. Vegetation- based remediation shows potential for accumulating, immobilizing, and transforming a low level of persistent contaminants. In natural ecosystems, plants act as filters and metabolize substances generated by nature. Phytoremediation is an emerging technology that uses plants to remove contaminants from soil and water.

The term "phytoremediation" is relatively new, coined in 1991. Its potential for encouraging the biodegradation of organic contaminants requires further research, although it may be a promising area for the future. We can find five types of phytoremediation techniques, classified based on the contaminant fate: phytoextraction, phytotransformation, phytostabilization, phytodegradation, rhizofiltration, even if a combination of these can be found in nature.

- *Phytoextraction* or *phytoaccumulation* is the process used by the plants to accumulate contaminants into the roots and aboveground shoots or leaves. This technique saves tremendous remediation cost by accumulating low levels of contaminants from a widespread area. Unlike the degradation mechanisms, this process produces a mass of plants and contaminants (usually metals) that can be transported for disposal or recycling.

- *Phytotransformation* or *phytodegradation* refers to the uptake of organic contaminants from soil, sediments, or water and, subsequently, their transformation to more stable, less toxic, or less mobile form. Metal chromium can be reduced from hexavalent to trivalent chromium, which is a less mobile and noncarcinogenic form.
- *Phytostabilization* is a technique in which plants reduce the mobility and migration of contaminated soil. Leachable constituents are adsorbed and bound into the plant structure so that they form a stable mass of plant from which the contaminants will not reenter the environment.
- *Phytodegradation* or *rhizodegradation* is the breakdown of contaminants through the activity existing in the rhizosphere. This activity is due to the presence of proteins and enzymes produced by the plants or by soil organisms such as bacteria, yeast, and fungi. Rhizodegradation is a symbiotic relationship that has evolved between plants and microbes. Plants provide nutrients necessary for the microbes to thrive, while microbes provide a healthier soil environment.
- *Rhizofiltration* is a water remediation technique that involves the uptake of contaminants by plant roots. Rhizofiltration is used to reduce contamination in natural wetlands and estuary areas.

In Table, we can see an overview of phytoremediation applications.

**Table. Overview of Phytoremediation Applications.**

| Technique | Plant mechanism | Surface medium |
|---|---|---|
| Phytoextraction | Uptake and concentration of metal viadirect uptake into the plant tissue with subsequent removal of the plants | Soils |
| Phytotransfo-rmation | Plant uptake and degrad-ation of organic compounds | Surface water,Ground Water |
| Phytostabilization | Root exudates cause metal to precipitate and become less available | Soils, groundwater, mine tailing |
| Phytodegradation | Enhances microbial degrada- | Soils,groundwater |

| | | |
|---|---|---|
| | tion in rhizosphere | within rhizosphere |
| Rhizofiltration | Uptake of metals into plant roots | Surface water and water pumped |
| Phytovolatilization | Plants evaportranspirate selenium, mercury, and volatile hydrocarbons | Surface Water and water pumped Soils and groundwater |
| Vegetative cap | Rainwater is evaportranspirated by plants to prevent leaching contaminants from disposal sites | Soils |

Phytoremediation is well suited for use at very large field sites where other methods of remediation are not cost effective or practicable; at sites with a low concentration of contaminants where only polish treatment is required over long periods of time; and in conjunction with other technologies where vegetation is used as a final cap and closure of the site.

There are some limitations to the technology that it is necessary to consider carefully before it is selected for site remediation: long duration of time for remediation, potential contamination of the vegetation and food chain, and difficulty establishing and maintaining vegetation at some sites with high toxic levels.

## BIOREMEDIATION: TECHNIQUES FOR CLEANING UP A MESS

Waste products resulting from human life have always been a serious problem. Today these waste products range from raw sewage to nuclear waste.

In the past disposal of these wastes meant digging a hole, dumping the waste material in, then filling it all in. Out of sight, out of mind. But lately this method has become insufficient. The toxic materials from these "dig and dump" sites have begun to leak into water sources and into areas that sustain human life. This problem has led to modern-day *bioremediation*.

Bioremediation is the transformation or degradation of contaminants into non-hazardous or less hazardouschemicals. Bacteria are generally used for bioremediation, but fungi, algae and plants have also been used. Bioremediation is not a new technology. There has been evidence that compost piles existed as far back as 6000 BC, and in 1891 the first biological sewage treatment plant was created in Sussex, UK. However, the word "bioremediation" did not appear in peer-reviewed scientific literature until 1987.

There are three classifications of bioremediation:·

- *Biotransformation*: The alteration of contaminant molecules into less or non-hazardous molecules·
- *Biodegradation*: The breakdown of organic substances in smaller organic or inorganic molecules·
- *Mineralization*: Is the complete biodegradation of organic materials into inorganic constituents such as $CO_2$ or $H_2O_4$.

These three classifications of bioremediation can occur either *in situ* (at the site of contamination) or *ex situ* (contaminant taken out of the site of contamination and treated elsewhere).There are advantages and disadvantages to both in situ and ex situ strategies. Ex situ strategies (aka "pump and treat") removes the contaminants and places them in a contained environment. This contained environment allows for easier monitoring and maintaining of conditions and progress, thus making the actual bioremediation process faster. However, the removal of the contaminant from the contaminated site is time consuming, costly and potentially dangerous.

By bringing the contaminants to the surface, the workers and the general public have increased exposure to the toxic material. There are several extraction strategies to facilitate ex situ bioremediation. The soil can actually be dug up and transported to a bioreactor. Soil washing is another method that can be used, where water is flushed through the contaminated region and then transferred to a bioreactor for treatment1. Similarly, soil ventingcan be used, where air is flushed through the contaminated region and the air containing the contaminant is transferred to a bioreactor for treatment.

The method of contaminant extraction depends on the nature of the contaminant in question(whether it is gas, liquid or solid phase, its chemical properties, and its toxicity).In contrast, the in situ strategy does not require removal of the contaminant from the contaminated site. Instead either *biostimulation* or *bioaugmentation* is applied. Biostimulation is the addition of nutrients, oxygen or other electron donors and acceptors to the coordinated site in order to increase the population or activity of naturally occurring microorganisms available for bioremediation.

Bioaugmentation is the addition of microorganisms that can bio-transform or biodegrade contaminants. The microorganisms added can be a completely new species or more members of a species that already exists at the site. The advantage of in situ bioremediation is that there is no need to extract the contaminants, so there is less exposure to workers, and it is also less costly. However, there are also disadvantages to this strategy. The site of bioremediation is not contained, therefore, it is harder to control conditions and monitor progress.

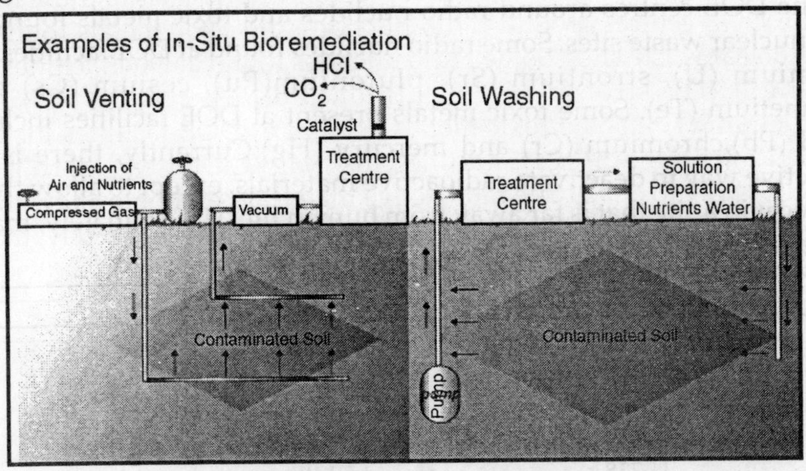


**Fig.** In-situ Bioremediation.

One example of this is the attempt to biostimulate microbes at an oil spill site. Nutrients added to the site end up diffusing through the water, and the result is a solute concentration the same as before the supplementary nutrients were never added5. Another disadvantage of in situ bioremediation is that if the soil(or other media that contains contamination) is heterogeneous, there will be uneven flow of liquid or gas containing the nutrients or microbes, so different are as will undergo different rates of remediation.

Despite the complications with bioremediation, it is still being used or studied for use in the remediation of crude oil spills, sewage effluents, chlorinated solvents, pesticides, agricultural chemicals, gasoline contaminants, contaminants from wood processing, radioactive metals and toxic metals.For the last 50 years, the United States have used nuclear energy for many purposes. However, there search, development and production of nuclear material has left the United States with a lot of nuclear waste to deal with. Normally this waste is just contained and buried.

This has resulted in more than 120 sites in 36 states and territories that contain hazardous and radioactive wastes, including 475 billion gallons of contaminated groundwater, 75 million cubic metres of contaminated sediment, and 3 million cubic metres of leaking waste buried in landfills, trenches and spill areas. In the early 1990's the threat of the Cold War ended and the US Department of Energy(DOE) shifted their focus towards cleaning up these sites. The first few years consisted mainly of cataloging and preliminary characterization, which cost more than$23 billion (the next ten years are budgeted to exceed$60 billion).

Such a large and costly project must have priorities and the focus of the DOE centres around radio nuclides and toxic metals found in the nuclear waste sites. Some radio nuclides found at DOEfacilities are uranium (U), strontium (Sr), plutonium(Pu), cesium (Cs) and technetium (Te). Some toxic metals present at DOE facilities include lead (Pb),chromium (Cr) and mercury (Hg).Currently, there is no effective way to deactivate radioactive materials, except to allow them to decayin a site that is far away from human life. Unfortunately, many of radio nuclides have very long half-lives:

| Radionuclide | Half-life |
|---|---|
| Sr-90 | 28 years |
| Cs-137 | 30 years |
| Pu-239 24, | 100 years |
| Tc-97 | 2.6 million years |
| U-238 | 4.5 billion years |
| U-235 | 7.13 million years |

Exposure to this radiation can cause radiation burn sand cancer. In addition, strontium (Sr) can replace calcium in bones and become concentrated enough to cause radiation injury from within the body. The problem with containment sites of these radio active materials is that they can leak. Most of the leaking is caused by solublization of the metals into water. This allows the radioactive metals to travel with the water, which may flow through or around the site. One solution against this problem is to make the metals insoluble.

This essentially makes the radio nuclides immobile and prevents leaking from the contaminated sites. One way of doing this is to introduce metal reducing bacteria to the containment site and allowing the bacteria to precipitate the metals out of the solution. The bacteria can directly reduce the radioactive metals from an oxidized, soluble form to a reduced, insoluble form. The bacterium does this by taking electrons from organic compounds, and using the radioactive metal as the final electron acceptor in the process of energy production. For example, a bacterium can reduce U (VI), Cr (VI) and Tc (VII)to U (IV), Cr (III) and Tc (IV).

The bacteria can also indirectly reduce the radioactive metals by using an intermediate electrondonor. For example, the bacterium can use either$Fe^{3+}$ or $SO_4^{2-}$ as a final electron acceptor. $Fe^{3+}$becomes $Fe^{2+}$ and $SO_4^{2-}$ becomes $H_2S^-$ $Fe^{2+}$ and$H_2S^-$ can then reduce the radioactive metals to an insoluble state. The result of these redox reactions can be seen as the precipitant within the metal-reducing bacteria. Radioactive metals are not the only contaminants on which this redox bioremediation technique can be used. It can also be used to immobilize toxic metals or change them into less toxic forms. The negative effects of a few toxic metals include:

- Cr (IV)- Throat and lung cancer, shortenedlifespan, reproductive problemsand lower fertility.
- Pb - Plumbism (lead poisoning),anaemia, effects on the intestines and central nervous system. Alsoin children (generally having a less well-developed blood brain barrier than adults): behavioural changes; decreased intelligence; brain damage and even death have been observed.
- $CH_3Hg$- Neurotoxicity.

Although using bioremediation is a great idea, quite often the contaminants or the contents of the contaminated site are also toxic to the active microbe involved in the bioremediation process. This problem can make it very difficult to keep the rate of bioremediation high. A solution to this problem is genetically engineered microbes (GEM) that are resistant to the extreme conditions of the contaminated site and also have bioremediary properties. For example, radiation is damaging to most life forms including most bacteria.

However, there is one species of bacterium, *Deinococcus radiodurans,* that is resistant to radiation damage (it can withstand up to1.5 million rads of radiation. If the genes for metal reduction can be transferred into this radiation-resistant species, or the radiation-resistant genes transferred into the metal-reducing bacteria, a GEM could be made that would be more efficient at bioremediation of sites leaking radioactive metals. This is only one exampleof bacterial species being studied for their ability to resist extreme conditions. Others include *bacillusinfernos,* which can resist high temperatures, and *Methanococcus jannaschii,* which can resist high pressures (up to 230 atm) in addition to high temperatures.

*Dehalococcoides* have the natural ability to dechlorinate tri-, tetra- and penta-chloro dioxins. The genes responsible for this dechlorination have been identified and modified to become over expressed. This allows the dechlorination abilities of *Dehalococcoides*to be augmented. Further studies into placing theseoverexpressed genes into more robust organisms for bioremediation purposes are being done. The value of a GEM with the ability to bioremediate sites with dioxins is obvious when considering the toxicity of dioxins. Acute dioxin toxicity causes symptoms such as disturbance of hair arrangement, atrophy of the thymus gland, and hypertrophy of the liver (according to animal tests) and a skin disease named chloracne in humans.

Chronic poisoning include scarcino genicity, liver disorders, teratogenicity, and heterotopic endometriosis. Dioxins are also suspected of causing endocrine disruption. Intake of dioxins over the tolerable daily limit can result in degradation of immunity, malignant tumours and morphologicalde formations of reproductive organs. Sources of dioxins included waste incinerators, impurities in agricultural chemicals and products containing PCBs. Dioxins can be found in air, water, soil, food and even breast milk.

Microbes are not the only species that can been hanced by genetic modification for bioremediatory purposes. Plants have also been studied and used. Bioremediation by plants is called *phytoremediation*. Arsenic is one target of phytoremediation. The health effects of arsenic include liver, lung, kidney and bladder cancers. One plant, *Arabidopsis thaliana* has been genetically modified to over express two bacterial genes, *arsC* and g-*ECS*. The gene *arsC* codes for arsenate reductase, which allows the plant to modify arsenate into are·senite, and g-*ECS* codes for gglutamylcysteine synthase, which makes a thiol that can detoxify aresenite by forming arsenic-protein thiates that are then stored in vacuoles. Essentially this genetically modified plant can take up arsenate, detoxify it and store it.

Phytoremediation can also beused to destroy high-energy compounds such as TNT,GTN, RDX, TETRYL and HMX. Tobacco plants have been genetically modified to express bacterial pentaerythritol tetanite (PETN) reductase allowing these plants to take up high-energy compounds and reduce them to non-explosive substances. Another genetically modified plant possesses a bacterial mercuric reductase gene allowing it to take up mercury(Hg) out of the soil and store it safely.Although genetically modified plants and GEMs have great applications in bioremediation, creating them has serious ethical implications. For this reason, the DOE has formed a group called BASIC(Bioremediation And its Societal Implication and Concerns) Programme.

This group consists of community leaders, engineers, representatives, scientists and lawyers. Together they discuss the ethical and societal implications of bioremediation, foster respect and collaboration between scientist and stock holders, and enhance research communication and develop educational materials. One of the largest concerns with GEMs is that they are not naturally occurring and therefore they could potentially upset the natural environment. This concern can be extended to bioremediation that does not even use GEMs: in bio augmentation, the addition of more microbes to a naturally occurring population in the contaminated site can upset its natural balance. Further more, the addition of a new species of microbe to the site can also upset the balance of co-existing microbe populations present.

In biostimulation, the preexisting microbe population responsible for bioremediation is being expanded beyond its normal size and this could result in a disturbance to the balance of surrounding microbe populations. Also, the nutrients added for biostimulation could stimulate microbe species other than that intended for bioremediation. The consequences of this are unknown. Although ideally the microbes added for bioaugmentation or expanded by biostimulation would die off when the addition of nutrients ceased, there is the possibility that these expanded populations would continue to thrive, resulting in severe consequences to the environment.

Especially with in situ bioremediation, where the contamination site is not remediated within a contained bioreactor, these environmental balance issues must be considered and addressed. The balance between the damage to the environment caused by toxic materials and the potential damage that may be caused by bioremediating microbes must be determined. Bioethics concerning GEMs and ecological damage control are not the only factors outside of science that need to be considered in bioremediation. Cost is also a important factor. How much is it worth in dollars to clean up a contaminated site?

It has been predicted that by 2010, the global environmental technology industry could be worth more than $1.5trillion, of which biotechnological processes could account for 15-20%. This large quantity of money has not been overlooked by businesses. Many bioremediation and related technologies companies have formed: Battelle Environmental Systems &Technology, Bioremediation Service, Inc., BioSolutions International, Inc., Chemical Specialities International, International Biochemicals (InterBio),Micro-Bac International, Inc., Oppenheimer Biotechnology, Inc. (OBI), Phytotech, Inc., Regenesis Bioremediation Products, Inc Even though bioremediation reduces the cost of decontaminating sites significantly, it still costs money and time. Butultimately, the question is not whether decontaminationis worth the effort and expense. The question is whether or not we can afford to neglect these environmental problems that have accumulated overtime.

## BIOREMEDIATION AS AN EMERGING TECHNOLOGY FOR THE DETOXIFICATION OF OBSOLETE PESTICIDES

Bioremediation as defined by the American Academy of

Microbiology is "the use of living organisms to reduce or eliminate environmental hazards resulting from accumulations of toxic chemicals or other hazardous wastes". Bioremediation is not a new technology. This is evident in that humankind has practiced composting, sewage treatment and fermentation since the beginning of recorded history.

All of these processes utilize microbial processes in degradation process. The modern use of bioremediation began with the opening of the first biological sewage treatment plant in Sussex, UK, in 1891. The use of this technology in cleaning up pollutant spills is gaining popularity. Over the past ten years an increase in the types of contaminants to which bioremediation is applied has been evident.

Indigenous and enhanced micro-organisms have been shown to degrade industrial solvents, polychlorinated biphenyls, explosives and many different agricultural chemicals, including pesticides. Pilot, demonstration, and full-scale applications of bioremediation have been carried out, but on a limited scale. Internationally micro-organisms have been identified and employed, to transform and degrade contaminants. Admittedly the application of this technology to *in situ* bioremediation of polluted sites has been limited.

At present, bioremediation is often the preferred method for remediation of especially petroleum hydrocarbons, because it is cost effective, and it converts the petroleum hydrocarbons into the harmless by-products carbon dioxide and water. Bioremediation can be used to degrade concentrated organic contaminants near their source or as a secondary remediation strategy following specific physical or chemical methods. It can also be applied for sequestration of metals and radionuclides through microbially mediated transformation processes as well as for remediation of large plumes of diluted contaminants that are broadly dispersed in the environment.

However, the full benefits of bioremediation have not been realized because processes and organisms that are effective in controlled laboratory tests are not always equally effective in full-scale applications. The failure to perform optimally in the field setting stems from a lack of predictability due, in part, to inadequacies in the fundamental scientific understanding of how and why these bioremediation processes work.

Micro-organisms degrade or transform contaminants by a variety of mechanisms. Petroleum hydrocarbons for example are converted to carbon dioxide and water or are used as a primary food source by bacteria, which use the energy to generate new cells. Where the hydrocarbons are chlorinated the degradation takes place as a secondary or co-metabolic process rather than a primary metabolic process. In such a case enzymes, which are produced during aerobic utilization of carbon sources such as methane, degrade the chlorinated compounds. Under aerobic conditions, a chlorinated solvent such as trichloroethylene can be degraded through a sequence of metabolic steps, where some of the intermediary by-products may be more hazardous than the parent compound (e.g., vinyl chloride).

The impressive capabilities of micro-organisms and plants to degrade and transform contaminants should provide tremendous benefits in the clean-up of pollutants from spills and storage sites. These remediation ideas have provided the foundation for many *ex situ* waste treatment processes (including sewage treatment) and a host of *in situ* bioremediation methods that are in practice today.

Over the past ten years, progress has been made in expanding the number and type of contaminants to which bioremediation can be applied and in the number of practical methods for implementing *in situ* bioremediation. Techniques such as hydrofracturing have been developed for improved delivery of nutrients to micro-organisms in low-permeability geologic media. Methods have been developed for creating passive treatment systems such a bio-filters.

Novel concepts for using microbial produced biopolymers as *in situ* plugging agents have also been explored. Bioremediation is a more cost effective method of remediation as compared to incineration or physical and chemical remediation methods. This technology has the potential to be one of the most cost effective technologies for dealing with environmental remediation problems.

## IN SITU OR EX SITU

*In situ* bioremediation refers to below ground methods applied at the site of contamination whereas *ex situ* refers to above-ground bioremediation, where the sediment or water has been extracted from the subsurface. The most widespread use of ex situ bioremediation is the cleanup of storage tank leaks and oil spills in pipelines, tank farms and petroleum refineries. The technique, known as a solid-phase bioremediation is similar to composting, in which soils are physically and chemically manipulated to stimulate breakdown of target organics by resident and exogenous microbes. In soil piles it is necessary to install piping networks to distribute air and allow addition of water and supplemental additives.

Application of nutrients, pH regulating chemicals, and microbial strains may be performed by spraying (in water slurry) or direct application with tillage to distribute the additives into the soil. For more heavily contaminated soils and sludge's, slurry processes are used.

These are modified versions of the activated sludge process used in various industrial wastewater treatment plants. The soil or sludge is fluidised with water and treated in an on site bioreactor. These systems are easily monitored and addition of nutrients and microbial strains can be made to closely control and enhance the bioprocess.

The third major category of on site bioremediation involves pumping groundwater to the surface for treatment in an above-ground bioreactor. The effluent from the bioreactor, containing oxygen, nutrients and acclimated micro-organisms, is then injected back into the ground to remediate the contaminated soils associated with the groundwater. This process combines *ex situ* groundwater treatment with *in situ* soil treatment.

*In situ bioremediation has several advantages over ex situ techniques*:

- In situ bioremediation can be used to completely degrade and detoxify some organic contaminants, thereby permanently removing liability for the contaminants.
- For deep, widely dispersed plumes of heavy metals and radionuclides, in situ bioremediation and immobilization may be the only viable solution.
- For some types of contaminants, physical and chemical methods of remediation may not completely remove the contaminants, leaving residual concentrations that are above regulatory guidelines. Bioremediation can be used as a cost-effective secondary treatment scheme to decrease the concentration of contaminants to acceptable levels. In other cases, bioremediation can be the primary treatment method, and followed by physical or chemical methods for final site closure.
- In some cases, natural attenuation (including natural bioremediation) of the contaminant plumes may be the only cost effective solution.
- For complex mixtures of contaminants requiring a combination or sequence of physical and chemical remediation methods, bioremediation techniques that use microbial consortia to concurrently address all contaminants may be faster and more cost-effective.

## Biodegradation of Pesticides

Microbial degradation is an important step in the disappearance and, in most cases detoxification of pesticides. Herbicide biodegradation may prevent the problem of environmental pollution but it can also reduce the effectiveness of a compound in controlling targeted pests. Many soil applied pesticides are degraded more rapidly following repeated application at the same site.

In a survey of soils from commercial fields, there was evidence that enhanced biodegradation of the compound has been induced by normal field applications, in some soils by a single previous treatment. Several herbicides are prone to degradation, including members of the thiocarbamate group, the ureas, linuron and monolinuron, the amides, propyzamide and napropamide, and the triazinones, chloridazon and metamitron.

Factors responsible for the enhanced degradation are micro-organisms present in the soil, able to degrade applied pesticide. Walker *et al.*, 1993, gave evidence, that bacteria are responsible for the process. The rate of degradation in soil was unaffected by treatment of the soils with the antifungal antibiotic cycloheximide, but was inhibited by the antibacterial antibiotic chloramphenicol. Usually mixed cultures of bacteria able to degrade pesticides are isolated from these soils by enrichment culture.

Early research indicated that napropamide was relatively stable to degradation in the soil, but the results of recent laboratory and field experiments have indicated that the rate of degradation is increased significantly in the soil treated previously with the same compound. Mixed cultures of bacteria were isolated from the soils by enrichment culture on a liquid mineral medium with napropamide as a sole carbon and nitrogen source. The related herbicide diphenamid is also susceptible to enhanced degradation, suggesting that substituted amides may be particularly prone to this phenomenon.

Enhanced degradation of linuron occurred in soils, following repeated applications of the herbicide. These effects were more pronounced in laboratory incubations than in field persistence studies. Direct isolation of bacteria from soil failed to identify bacteria with the ability to degrade the herbicide. Sequential enrichment culture in a mineral base medium led to the isolation of an apparently stable mixed culture of bacteria, which would degrade the herbicide.

Metamitron is a selective, pre and post-emergence herbicide, used for weed control in sugar beet, which acts by inhibiting electron transport in photosynthesis. The compound is degraded to desaminometamitron in various soils and by different soil microbes. Parekh *et al.*, 1994, isolated metamitron-degrading bacteria *Rhodococcus* sp. from the treated soils by enrichment culture. The herbicide was completely degraded within 7 days at 25° C in laboratory tests. A pure culture of *Arthrobacter* sp. was shown to degrade metamitron, in the presence of alternative sources of carbon and nitrogen, within 2 weeks of incubation in the dark. Bacteria capable of degrading carbofuran as sole carbon and nitrogen sources were isolated from liquid cultures of treated soils. Similar types of carbofuran-degrading bacteria were isolated from different soils. All isolates were Gram negative, aerobic rods that hydrolyzed the insecticide to carbofuran phenol. Racke and Coats, 1990, imply the presence of carbofuring degrading bacteria in soils without a history of previous treatment.

It is possible that many soils contain a small population of bacteria that have the ability to degrade certain compounds. The presence of such bacteria in untreated soils could be due to contamination from surrounding treated areas or they may be an inherent part of the soil microflora, which degrade naturally occurring analogous compounds. Chary *et al.*, 1992, detected carbofuran degrading bacteria in 48% of samples from previously untreated plots.

Walker *et al.*, 1993, isolated a mixed microbial culture able to degrade the herbicide to a single degradation product – napthoxypropionic acid. Addition of this culture to a previously untreated soil introduced rapid degrading ability. These results gave hope for use of augmentation for the treatment of sites contaminated by carbofuran and other pesticides.

Atrazine is a broad-leaf, pre-emergence herbicide. It is a leading member of a class of triazine ring-containing herbicides that includes simazine and terbuthylazine.

Atrazine has been found to be less biodegradable than other pesticides. However, a number of different bacteria have been identified that are capable of metabolizing atrazine to ammonia and carbon dioxide.. Organisms that can initiate the pathway are given, but other organisms may also carry out later steps. This pathway demonstrates microbial biodegrative diversity.

With some chemicals in some soils the waste is consumed completely, or has been reduced to an acceptable concentration level. In other kinds of soils, or when the toxic agents have low water solubility, or have been adsorbed in the soil matrix to a high degree, the micro-organisms would not be able to achieve sufficient degradation.

In soils contaminated with pesticides, a unique problem exists because of the presence of herbicides; chemicals designed to prohibit the growth of vegetation. Herbicide tolerant plants can survive at these sites and are ideal candidates for testing whether vegetation can be used to enhance microbial degradation of pesticide wastes.

Advantages

- The need for trained personnel at local level allows for

potential development of small to medium scale enterprises, which could aid in job creation in especially rural areas. The technology required to multiply micro-organisms may be ideally suited for small business development. In addition the technology developed will remain in the country and continent.

- The technology has a high potential for technology transfer
- Non-disturbance of the soil will insure soil integrity and fertility remains intact.
- The implementation of a bioremediation system will require labour, which can be provided through training of locals.
- The implementation of the system is likely to be more cost effective than incineration, especially where soil contamination is addressed.

*Disadvantages*

- The main disadvantage of implementation of bioremediation systems as a destruction technology lies in the longer time required for clean-up. The sites to be addressed are likely to contain a mixture of contaminants. A single micro-organisms system is not likely to degrade all the pollutants components. It will likely be necessary to isolate and culture stocks of micro-organisms for some pollutants. There is thus likely to be some development phase required prior to implementation.
- The implementation of such a project will require training of local personnel to enable them to apply all required aspects of the technology.
- Bioremediation requires a longer term monitoring action to be implemented through which the pollution status must be determined
- Some compounds may not be prone to microbial degradation, and may require an additional chemical degradation step or bio-augmentation step.

# Index